The Purchase of the North Pole

Topsy Turvy

By
Jules Verne

18
EasyRead Super Large

ReadHowYouWant Classics Library

Set in 18 pt. TiresiasLPfont

TABLE OF CONTENTS

CHAPTER I

IN WHICH THE NORTH POLAR PRACTICAL ASSOCIATION *RUSHES A DOCUMENT ACROSS TWO WORLDS*

***

“Then Mr. Maston, you pretend that a woman has never been able to make mathematical or experimental-science progress?”

“To my extreme regret, I am obliged to, Mrs. Scorbitt,” answered J.T. Maston.

“That there have been some very remarkable women in mathematics, especially in Russia, I fully and willingly agree with you. But, with her cerebral conformation, she cannot become an Archimedes, much less a Newton.”

“Oh, Mr. Maston, allow me to protest in the name of my sex.”

“A sex, Mrs. Scorbitt, much too charming to give itself up to the higher studies.”

“Well then, according to your opinion, no woman seeing an apple fall could have discovered the law of universal gravitation, so that it would have made her the most illustrious scientific person of the seventeenth century?”

“In seeing an apple fall, Mrs. Scorbitt, a woman would have but the single idea – to eat it – for example, our mother Eve.”

“Pshaw, I see very well that you deny us all aptitude for high speculations.”

“All aptitude? No, Mrs. Scorbitt, and in the meanwhile I would like to prove to you that since there are inhabitants on earth, and consequently women, there has not one feminine brain been found yet to which we owe any discoveries like those of Aristotle, Euclid, Kepler, Laplace, etc.”

“Is this a reason? And does the past always prove the future?”

"Well, a person who has done nothing in a thousand years, without a doubt, never will do anything."

"I see now that I have to take our part, Mr. Maston, and that we are not worth much."

"In regard to being worth something" – began Mr. Maston, with as much politeness as he could command.

But Mrs. Evangelina Scorbitt, who was perfectly willing to be satisfied, answered promptly: "Each one has his or her lot in this world. You may remain the extraordinary calculator which you are, give yourself up entirely to the immense work to which your friends and yourself will devote their existence. I will be the woman in the case and bring to it my pecuniary assistance."

"And we will owe you an eternal gratitude," answered Mr. Maston.

Mrs. Evangelina Scorbitt blushed deliciously, for she felt, according to report, a singular sympathy for J.T. Maston. Besides, is not the heart of a woman an unfathomable gulf?

It was really an immense undertaking to which this rich American widow had resolved to devote large sums of money.

The scheme and its expected results, briefly outlined, were as follows:

The Arctic regions, accurately expressed, include according to Maltebrun, Roclus, Saint-Martin and other high authorities on geography:

> 1st. The northern Devon, including the ice-covered islands of Baffin's Sea and Lancaster Sound.
>
> 2d. The northern Georgia, made up of banks and numerous islands, such as the islands of Sabine, Byam-Martin, Griffith, Cornwallis, and Bathurst.
>
> 3d. The archipelago of Baffin-Parry, including different parts of the circumpolar continent, embracing Cumberland, Southampton, James-Sommerset, Boothia-Felix, Melville, and other

parts nearly unknown. Of this great area, crossed by the 78th parallel, there are over 1,400,000 square miles of land and over 700,000 square miles of water.

Within this area intrepid modern discoverers have advanced to the 84th-degree of latitude, reaching seacoasts lost behind the high chain of icebergs which may be called the Arctic Highlands, given names to capes, to mountains, to gulfs, to bays, etc. But beyond this 84th degree is mystery. It is the terra incognita of the chart-makers, and nobody knows as yet whether behind is hidden land or water for a distance of 6 degrees over impassable heaps of ice to the North Pole.

It was in the year 189– that the Government of the United States conceived the idea of putting the as yet undiscovered countries around the North Pole up at auction sale, and an American society had just been formed with the plan of purchasing this Arctic area and has asked the concession.

For several years, it is true, the Conference at Berlin had formulated a special plan for the guidance of such of the great powers as might wish to appropriate rights under the claim of colonization or the opening of commercial markets. This code was not acceptable to all, and the Polar region had remained without inhabitants. As that which belongs to none belongs to every one, the new Society did not wish merely to occupy it, but to purchase it outright, and so avoid further claims.

There never is in the United States any project so bold as not to find people to regard it as practical and back it with large amounts of money. This was well shown a few years ago when the Gun Club of Baltimore tried to send a projectile to the moon, hoping to obtain a direct communication with our satellite. Was it not enterprising Americans who furnished funds for this undertaking? Large amounts were necessary for this interesting trial and were promptly found. And, had it been realized, would we not have to thank the members of that club who had dared to take the risk of this super-human experience?

Should a Lesseps propose to dig a channel across Europe to Asia, from the banks of the Atlantic to the waters of China; should a well-sinker offer to bore from the curb-stones to reach the beds of molten silicates, to bring a supply to your fireplaces; should an enterprising electrician want to unite the scattered currents over the surface of the globe into one inexhaustible spring of heat and light; should a bold engineer conceive the idea of putting the excess of Summer temperature into large reservoirs for use during the Winter in our then frigid zones; should an anonymous society be founded to do any of a hundred different similar things, there would be found Americans ready to head the subscription lists and a regular stream of dollars would pour into the company safes as freely as the rivers of America flow into the ocean.

It is natural to expect that opinions were very varied when the news spread that the Arctic region was going to be sold at auction for the benefit of the highest and final bidder, particularly when no public subscription list was started in view of this purchase, as the capital had all been secured beforehand.

To use the Arctic region? Why, such an idea could “only be found in the brain of a fool,” was the general verdict.

Nothing, however, was more serious than this project. A prospectus was sent to the papers of the two continents, to the European publications, to the African, Oceanic, Asiatic, and at the same time to the American journals. The American newspaper announcement read as follows:

To the Inhabitants of the Globe:

“The Arctic region situated within the eighty-fourth degree could not heretofore have been sold at auction for the very excellent reason that it had not been discovered as yet.

“The extreme points reached by navigators of different countries are the following:

“82° 45' , reached by the English explorer, Parry, in July, 1847, on the twenty-eighth meridian, west, to the north of Spitzberg.

“83° 20' 28” , reached by Markham, with the English expedition of Sir John Georges Nares, in May, 1867, on the fiftieth meridian, west, in the north of Grinnell Land.

“83° 35' latitude, reached by Lockwood and Brainard, of the American expedition under Lieut. Greely, in May, 1882, on the forty-second meridian, west in the north of Nares Land.

“The property extending from the eighty-fourth parallel to the pole on a surface of six degrees must be considered an undivided domain among the different states of the globe and not liable to be transformed into private property through a public auction sale.

“No one is compelled to live in this section, and the United States, relying on this non-ownership, has resolved to provide for the settlement and use of the domain. A company has been founded at Baltimore under the name of the North Polar Practical Association, representing officially the American Union. This Company intends to purchase the said country according to the common law, which should then give them an absolute right of proprietorship to the continent, islands, inlets, waters, rivers, etc.; in fact, of everything of which the Arctic region is composed. It is well understood by the law of nations that this title of proprietorship cannot be touched under any circumstances, no matter what shall happen.

“These conditions having been laid before all the powers, the Arctic region is to be sold at public auction for the benefit of the highest

> and last bidder. The date of the sale is set for the 3d of December of the current year, in the Auction Hall at Baltimore, Maryland, United States of America.
>
> “Address for information Mr. W.S. Forster, Temporary Agent for the North Polar Practical Association, 93 High Street, Baltimore.”

The reader may imagine how this communication was received by the public at large. Most people considered it as an absurd idea. Some only saw in it a sample of characteristic American humbug. Others thought that the proposition deserved to be fairly considered, and they pointed to the fact that the newly-founded company did not in any way appeal to the public for pecuniary help, but was willing to do everything with its own capital. It was with its own money that it wanted to purchase the Arctic region. The promoters did not try to put gold, silver, and banknotes into their pockets and keep them for their own benefit. No, they only asked permission to pay for the land with their own money.

Some people who claimed to know said that the Company could have gone to work and taken possession of the country without any further ceremony, as it was their right as first occupants. But that is just where the difficulty came in, because until this time the Pole seemed to be forbidden ground to any one. Therefore, in case the United States should give possession of the country, the Company wanted a regular title to it without trouble about the matter in the future. It was unjust to blame them in any way, as in such an affair too many precautions cannot be taken. Besides, the circular had a paragraph which provided for all future chances. This paragraph was capable of so many interpretations that the exact meaning of it could not be rendered even by those who studied it closely. It was stipulated that the right of proprietorship should not depend upon any chances or changes in the country, no matter whether these changes were in the position or climate of the country.

What did this phrase mean? How could there ever be any changes in the geography or meteorology of a country like this one to be sold at auction? "Evidently," said some shrewd ones, "there must be something behind it."

The commentators had free swing and exercised it with a will. One paper in Philadelphia published the following pleasant notice:

"Undoubtedly the future purchasers of the Arctic region have information that a hard stone comet will strike this world under such conditions that its blow will produce geographic and meteorologic changes such as the purchasers of the Arctic region will profit by."

The idea of a blow with a hard stone planet was not accepted by serious people. In any case it was not likely that the would-be purchasers would have been informed of such a coming event.

"Perhaps," said a New Orleans newspaper, "the new Company thinks the precession of the equinox will in time favor the conditions likely to lead to the utilization of this domain."

"And why not? Because this movement modifies the direction of the axis of our spheroid," observed another correspondent.

"Really," answered the *Scientific Review*, of Paris. "As Adhemar has predicted in his book on the ocean currents, the precession of the equinox, combined with the movement of the earth's axis, will be such as to modify in a long period the average temperature of the different parts of the earth and in the quantities of ice accumulated around the two poles."

"This is not certain," replied the *Edinburgh Review*, "and, besides, supposing that this would be the case, is not a lapse of 12,000 years necessary before Vega becomes our polar star in consequence of this movement and the situation of the Arctic territory consequently changed in regard to its climate?"

"Well," said the Copenhagen *Dagblad*, "in 12,000 years it will be time to make preparations, and before that time risk nothing – not even a cent."

It was possible that the *Scientific Review* was right with Adhemar. It was also very probable that the North Polar Practical Association had never counted on this modification of climate due to the precession of the equinox. In fact, nobody had clearly discovered what this last paragraph in the circular meant or what kind of change it had in view.

Perhaps to know it, it would suffice to write to the Secretary of the new Society, or particularly its President. But the President was unknown. Unknown as much as the Secretary and all other members of the Council. It was not even known where the document came from. It was brought to the offices of the New York newspapers by a certain William S. Forster, a codfish dealer of Baltimore, a member of the house of Ardrinell & Co. Everything was so quiet and mysterious in the matter that the best reporters could not make out what it was all about. This North Polar Association had been so anonymous that it was impossible even to give it a definite name.

If, however, the promoters of this speculation persisted in making their *personnel* an absolute mystery, their intention was clearly indicated by the document spread before the public of two worlds.

Really, after all, the question was the purchase of that part of the arctic regions bounded by the 84th degree, and of which the North Pole was the central point. Nothing very exact concerning this region was known. The modern discoverers who had been nearest to this parallel were Parry,

Markham, Lockwood and Brainard. In regard to the other navigators of the northern seas they stopped far short of the above-mentioned point – such as Payez, in 1874, to 82° 15' north of the land of Francis Joseph, of New Zemble; Leout, in 1870, to 72°74' above Siberia; De Long in the *Jeanette* expedition, in 1879, to 78° 45' around the islands which bear his name. Others went around New Siberia and Greenland to the end of the Cape Bismarck, but had not passed the 76th, 77th, or 78th degree of latitude. The North Polar Practical Association wanted then a country which had never been touched before by mankind or discoverers, and which was absolutely uninhabited.

The length of this portion of the globe surrounded by the 84th degree, extending from the 84th to the 90th, making six degrees, which at sixty miles each make a radius of 360 miles and a diameter of 720 miles. The circumference therefore is, 2,260 miles and the surface 407 [square] miles. This is about the tenth part of the whole of Europe. A very desirable slice of land indeed. The document, as we have seen, also stated that these regions were not yet known geographically, belonged to no one and therefore belonged to everyone. But it could be foreseen that the adjoining States at least would consider these regions as the prolongation of their own possession towards the north and would consequently claim the right of ownership. Their pretensions would have more justice than those of discoverers who operated upon the whole of the Arctic countries and made explorations only for the glory of their own nation. The Federal Government represented in the new Society intended to make their rights valuable and to indemnify them for the price of their purchase. However it was the partisans of the North Polar Practical Association did not announce; the proprietorship was clear, and nobody being compelled to live there could object to the auction sale of this vast domain.

The countries whose rights were absolutely established as much as those of any countries could be were six in number – America, England, Russia, Denmark, Sweden-Norway and Holland.

Other countries could claim discoveries made by their mariners and their travellers.

France could interfere because some of her children had taken part in the expeditions sent out to conquer the territories around the pole.

Among the others the courageous Bellot, who died in 1853, in the islands of Beechey, during the Phoenix Expedition sent in search of Sir John Franklin. Nor must one forget Dr. Octave Pavy, who died in 1884, near Cape Sabine, while the Greely Mission was at Fort Conger. And the expedition which, in 1838–39, had gone to the Sea of Spitzberg with Charles Marmier, Bravais and his courageous companions, would it not be unfair to forget them. But despite all this France did not care to interfere in this commercial rather than scientific matter, and she abandoned all her rights for a share of the polar pie. The same of Germany. It had sent since 1671 the Hamburg expedition of Frederic Martens to the Spitsbergen, and in 1869 and '70 the expeditions of the *Germania* and of the *Hansa,* commanded by Koldervey and Hegeman, which went as far as Cape Bismarck by going along the coast of Greenland. But even if they had made so many brilliant discoveries they did not care to add a piece of the polar empire to that of Germany. The same was true with Austria, which was already possessor of the land of Francis Joseph, situated north of Siberia.

In regard to Italy having no right to interfere, she did not interfere at all; which is as strange as it is true. Then, also, there were the Esquimaux, which are at home in those places, and the inhabitants of Greenland, of Labrador, of Baffin's Archipelago and of the Aleutian Islands, situated between Asia and America, and also the tribe of Tchouktchis, who inhabited the old Russian Alaska and who became Americans in 1867. These people, in reality the real aborigines, had nothing at all to say about the matter. And how could these poor wretches have said anything, as they did not even have any sum of money, no matter how small, with which to pay for the land which the North Polar Practical Association was going to buy. Perhaps they could have paid a small sum by giving skins, teeth or oil, and yet the land belonged to them more than to any others, as they were the first occupants of this domain which was going to be sold on auction. But the Esquimaux, the Tchouktchis, the Samoyedes were not consulted at all. So runs the world.

CHAPTER II

IN WHICH THE DELEGATES FROM ENGLAND*, HOLLAND, *SWEDEN, DENMARK AND RUSSIA ARE PRESENTED TO THE READER.

***

One thing was evident to the whole world at once, namely, that if the new association should succeed in buying the Arctic regions, those regions would become absolutely the property of America or rather of the United States, a country which was always trying to acquire something. This was not a pleasing prospect to rival governments, but nevertheless, as has been said, the different States of Europe and of Asia not neighboring to these regions, refused to take part in the proposed auction sale so long as its results seemed so problematical to them. Only the powers whose property touched the eighty-fourth degree resolved to make their rights valuable by the attendance of official delegates. That was all. They did not care to buy even at a relatively moderate price land the possession of which was only a possibility. In this as in all cases insatiable England gave orders to its financial agents to make an imposing showing. The cession of the polar countries did not threaten any European trouble nor any international complication. Herr von Bismarck, the grand Iron Chancellor, who was yet living, did not even knit his heavy brow. There remained only England, Denmark, Sweden, Norway, Holland and Russia to be present and make their bids to the Commissioner of Baltimore, against those of the United States.

It was a difficult matter to fix prices for this polar earth cap, the business value of which was at least very problematic. Their main reason for presenting themselves at the sale was that some advantage might accrue to them. Sweden and Norway, proprietors of the North Cape, situated beyond the seventy-second parallel, did not conceal the fact that they thought they had certain rights of proprietorship on these vast lands which extended to Spitsbergen, and from there to the North Pole. Denmark said that it had already in its possession islands and fiords on the line of the polar circle where their colonies had been founded, such as Disko Island, in the Davis Channel; the settlements of Holstein, of Proven, of Godhaven, of Uppernavik, in the Baffin Sea, and on the west coast of Greenland. Besides, did not the famous navigator, Behring (of Danish origin, although he was then in the service of Russia), in the year 1728 pass over the channel which afterwards carried his name before he started again, thirteen years later, and died miserably with thirty of his men on a little island, which also carries his distinguished name.

In the year 1619 did not the navigator, Jean Munk, explore the east coast of Greenland and discover several points formerly totally unknown? Therefore, Denmark had, she thought, undisputable rights to be proprietor of these regions.

In regard to Holland, there were her sailors Barentz and Heemskerk, who had visited the Spitsbergen and the New Zealand about the end of the sixteenth century. It was by one of her children too, Jean Mayen, through whose courageous campaign against the north the island which carries his name came in their possession. It is situated below the 72d degree of latitude. Therefore Holland thought her past had given her rights of possession. In regard to Russia, with Alexis Tschirikof, having Behring under his command;

with Paulutski, whose expedition advanced in 1751 beyond the limits of the ice-pack; with Capt. Martin Spangberg, and Lieut. William Walton, who dared to go into these unknown parts in 1739, she had taken a notable part in the search across the gulf which separates Asia and America.

Furthermore, the position of the Siberian territories, extending over 120 degrees to the extreme limits of Kamchatka, the length of the Asiatic coast, where the Samoyedes, Yakoutes, Tchuok-tchis, and other conquered people lived, did Russia not rule half of the Northern Ocean? And then, on the 75th parallel to within less than nine hundred miles from the pole, did she not possess the islands of the new Siberia, the Archipelago of Liatkow, discovered in the beginning of the eighteenth century? And finally, since 1764, before the English, before the Americans, before the Swedes, did not the navigator Tschitschagoff search a passage in the North to shorten the route between the two continents? However, notwithstanding this, it seemed that the Americans were more anxious to become possessors of this particularly inaccessible point of the globe than anyone else.

They had often tried to obtain it by devoting themselves to the search of Sir John Franklin, with Grinnel, with Kane, with Hayes, with Greely, with De Long, and other courageous navigators. They could also plead the geographical situation of their country, which develops itself below the polar circle from the Behring Sea to Hudson's Bay. And were not all these countries, all these islands – Wollaston, Prince Albert, Victoria, King William, Melville, Cockburne, Banks, Baffin, not counting the thousand small pieces of the archipelago – like a leaf spreading to the 90th degree? And then supposing that the North Pole should be attached by an uninterrupted line of territory to one of the large continents of the globe, would it not be more to America than to Asia or Europe? Therefore, nothing was more natural than the proposition to purchase this region by the Federal Government for the benefit of an American society.

If any power had undisputable modern rights to possess the polar domain it was certainly the United States of America. It must also be considered that the United Kingdom of Great Britain, which possessed Canada and British Columbia, numerous sailors of which had distinguished themselves in these Arctic countries, urged very good reasons for annexing this part of the globe to their vast empire. And its journals discussed the matter at great length. "Yes, without a doubt," answered the great English geographer, Kliptringan, in an article in a London newspaper, which made a great sensation; "yes, the Danes, the Hollanders, the Russians, and the Americans, can be proud of their rights." As for England, she did not wish to let this country escape her. Did not the northern part of the continent already belong to them? Have not these lands, these islands which composed them, been discovered and conquered by English discoverers since Willoughby, who visited Spitsbergen and New Zealand in 1739, to McClure, whose vessel made in 1853 the passage of the northwest? And then were not Frobisher, Davis, Hall, Weymouth, Hudson, Baffin, Cook, Ross, Parry, Bechey, Belcher, Franklin, Mulgrave, Scoresby, MacClinton, Kennedy, Nares, Collinson, Archer, all of Anglo-Saxon origin? And what country could make a more just claim on the portion of these Arctic regions that that which these navigators had been able to acquire? "Well," said a California journal, "let us put the matter on its real point, and as there is a question of *amour-propre* between the

United States and England, let us ask, If the English Markham of the Nares expedition had gone 83 degrees 20 minutes of latitude and the Americans, Lockwood and Brainard, of the Greely expedition, went to 83 degrees 35 minutes, to whom then does the honor belong of having come nearest to the North Pole?"

Such were the demands and explanations, but one could see that the struggle would only be active between American dollars and English pounds sterling. However, according to the proposition made by the North Polar Practical Association all countries had to be consulted and given a chance at the auction. The sale was announced to take place Dec. 3, at Baltimore. The sum realized by the sale was to be divided among the States which were unsuccessful bidders, and they were to accept it as indemnity and renounce all their rights in the Arctic regions for the future.

The delegates, furnished with their letters of credit, left London. The Hague, Stockholm, Copenhagen, and St. Petersburg, and arrived three weeks before the day fixed for the auction sale.

Up to this time America had only been represented by Mr. W.S. Forster, of the North Polar Practical Association.

The delegates of the European powers who had been chosen were included in the following list:

For Holland – Jacques Jansen, formerly Counsellor of the Netherlandish India; fifty-three years old, stout, short, well formed, small arms, small bent legs, round and florid face, gray hair; a worthy man, only a little incredulous on the subject of an undertaking the practical consequences of which he failed to see.

For Denmark – Eric Baldenak, ex-Sub-Governor of the Greenlandish possessions; of medium height, a little bent over, large and round head, so short-sighted that the point of his nose would touch his books; not willing to listen to any claim denying the rights of his country, which he considered the legitimate proprietor of the northern region.

For the Swedish-Norwegian peninsula – Jan Harald, Professor of Cosmography in Christiania; a genuine Northern man, red-faced, beard and hair blond; he regarded it as an established fact that the Polar region, being only occupied by the Paleocristic Sea, had absolutely no value. He was, however, not much interested in the matter and went there only as a duty.

For Russia – Col. Boris Karkof, semi-military man, semi-diplomat; a stiff, stubby mustache, seeming uncomfortable in his citizen clothes and feeling absent-mindedly for his sword which he was accustomed to carry; very much puzzled to know what was hidden in the proposition of the North Polar Practical Association, and whether it would not be the cause of international difficulties.

Finally for England – Major Donellan and his secretary, Dean Toodrink. The last two named represented all the tastes and aspirations of the United Kingdom, its commercial and industrial instincts, its aptitude to consider, by a law of nature, the northern regions their own property just as any country which did not belong to anyone else.

If there ever was an Englishman it was Major Donellan, tall, meagre, bony, nervous, angular, with a little cough, a head a la Palmerston, on bending shoulders; legs well formed after his sixty years; indefatigable, a quality he had well shown when he worked on the frontiers of India. He never laughed in those days, and perhaps never had. And why should he? Did you ever see a locomotive or a steam-engine or an elevator laugh? On this point the Major was very much different from his secretary, Dean Toodrink, a talkative fellow, very pleasant, with large head, and his hair falling on his forehead, and small eyes. He became well known on account of his happy manner and his taste for fairy tales. But, even if he was cheerful, he did not seem any less personally conceited than Major Donellan when he talked about Great Britain.

These two delegates were probably going to be the most desperate opponents to the American Society. The North Pole belonged to them; it always belonged to them. It was to them as if the Lord had given the mission to the English people to keep up the rotation of the earth around its axis, and as if it was their duty to prevent it passing into strange hands. It is necessary to observe here that France did not consider it necessary to send a delegate, but an engineer, of France, was present at the sale, just for the fun of it. We shall introduce him later on. The delegates of the Northern European States had arrived in Baltimore on different steamers, to give it the appearance that they had nothing at all to do with each other. They were really rivals. Each one of them had in his pocket the necessary means to fight against the American Society. But they could not fight with equal force. One could dispose of a sum of money which amounted to nearly a million, another could pass that amount. And really to purchase a piece of our globe to reach which seemed an impossibility, this ought still appear to be dear. In reality the best provided for was the English delegate, to whose order the Government had opened a very large credit. Thanks to this credit Major Donellan would not have very hard work to conquer his adversaries of Sweden, Denmark-Holland, and Russia. In regard to America – well, that was a different thing. It would be much more difficult to win against the fusillade of dollars. At least it was very probable that the mysterious society must have enough money on hand to go on in their work. Therefore, the highest bidding, which might come to millions, was between America and England.

As soon as the European delegates had landed public opinion became more excited. The most singular stories were printed in the newspapers. False theories were established, based on the purchase of the North Pole. What was the Society going to do with it? And what could they do with it? Nothing; or perhaps it was all done to corner the iceberg market. There was even a journal in Paris, the *Figaro,* which upheld this curious idea. But for this it would be necessary to pass south of the eighty-fourth parallel.

Be it as it may, however, the delegates who had avoided each other during their passage over the Atlantic became more and more associated after having arrived in Baltimore. Here is the reason: Since his arrival each one had tried to open communications with the North Polar Practical Association separately, unknown to the other. That which they wished to know were the motives hidden at the bottom of this affair and what profit the Society hoped to make out of the sale. Now, until the present time nothing indicated that the Society had opened an office at Baltimore. No office, no employees. All that could be learned was, "For information address only William S. Forster, High Street, Baltimore." And it did not look as though the honest consignee of codfish knew any more in this respect than the lowest street porter of the city.

The delegates could, therefore, learn nothing from him. They were accordingly compelled to rely upon the more or less absurd guesses of the public at large. Was the secret of the Society going to be kept inpenetrable as long as it did not make it known itself? This was the question. Without doubt it did not seem inclined to give any information on the subject until the purchase had been made. Therefore, it came that the delegates finished by seeing and meeting each other; they made visits to each other, and finally came in close communication with each other, perhaps with the idea of making a front against the common enemy, or, otherwise, the American Company. And so it happened that one evening they were all together in the Hotel Wolesley, in the suite occupied by Major Donellan and his secretary, Dean Toodrink, holding a conference. In fact, this tendency to a common understanding was principally due to the advice of Col. Boris Karkof, the best diplomat known. At first the conversation was directed to the commercial and industrial consequences which the Society pretended to gain by purchase of the Arctic domain. Prof. Jan Harald asked if any one had been able to gain any information on that point. All finally agreed that they had tried to get information from Mr. William S. Forster, to whom all letters should be addressed.

"I have failed," said Eric Baldenak.

"And I have not succeeded," added Jacques Jansen.

"In regard to myself," answered Dean Toodrink, "when I presented myself at the stores in High Street in the name of Major Donellan I found a large man in black clothes, wearing a high hat, with a white apron, which was short enough to show his high boots. When I asked him for information in the matter he informed me that the South Star had arrived with a full cargo from Newfoundland and that he was ready to furnish me with a fresh stock of codfish on account of Messrs. Ardronell & Co."

"And," answered the former counsellor of the Dutch Indies, always a little sceptical, "it would be much better to buy a load of codfish than to throw one's money into the ice-water of the North."

“This is not at all the question,” says Major Donellan, with a short and high voice. “The question is not the codfish, but the Polar region.”

“Americans ought to stand on their heads,” said Dean Toodrink, laughing at his own remark. “That will make them catch cold,” finally said Col. Karkof. “The question is not there,” said Major Donellan. “One thing only is certain, that for some reason or another America, represented by the N.P.P.A. (remark the word 'practical') wants to buy a surface of 407 square miles around the North Pole, a surface which is actually (remark the word 'actually') pierced by the eighty-fourth degree of latitude.”

“We know it, Major Donellan, and much more,” said Jan Harald. “But what we do not know is how the said company will make use of those countries or waters, if they are waters, from a commercial standpoint.”

“This is not the question,” answered for the third time Major Donellan. “A power wants to purchase with money a large part of the globe which, by its geographical situation, seems to belong especially to England” – “to Russia,” said Col. Karkof; “to

Holland," said Jacques Jansen; "to Sweden-Norway," said Jan Harald; "to Denmark," said Eric Baldenak.

The five delegates jumped to their feet, and it seemed as if the Council would turn to harsh words, when Dean Toodrink tried to interfere the first time. "Gentlemen," said he, in a tone of reconciliation, "this is not the question, following the expression of my chief," of which he made such frequent use. "As long as it has been decided that the Northern regions are going to be sold at auction, they will naturally belong to such representative who will make the highest bid for same. As long as Sweden, Norway, Russia, Denmark, Holland, and England have given large credits to their delegates, would it not be best for these nations to form a syndicate and raise a sum of money against which America could not make a bid? The delegates looked at each other. It was possible that Dean Toodrink had found the missing link. A syndicate – at present it is heard everywhere. Everything is syndicate nowadays, what one drinks, what one eats, what one reads, what one sleeps on. Nothing is more modern, in politics as well as business, than a trust. But an objection was started, or rather an explanation was needed, and Jacques Jansen tried to find out the sentiments of his colleagues by saying, “and afterwards,” yes, after the purchase of the region by the syndicate, then what? “But it seems to me that England,” said the Major in a rough voice, “and Russia,” said the Colonel, with nostrils terribly dilated, “and Holland,” said the Counsellor; “as God has given Denmark to the Danish,” observed Eric Baldenak – “Excuse me, there is only one country,” interrupted Dean Toodrink, “which has been given by our Lord, and that is the world.” “And why,” said the Swedish delegate? “Did not the poet say

'Deus nobis haec otia fecit,'”

said this merryman in translating according to his fashion the close of the sixth verse of the first eclogue of Virgil. All began to laugh except Major Donellan, who stopped for the second time the discussion which threatened to finish badly. Then Dean Toodrink said, “Do not quarrel, gentlemen. What good will it do us? Let us rather form a syndicate.”

“And afterwards?” asked Jan Harald.

“Afterwards,” answered Dean Toodrink, “nothing more simple, gentlemen. After you shall have bought the polar domain it will remain undivided among us or will be divided after a regular indemnity to one of the States which have been purchasers. But our purpose would have already been obtained, which is to save it from the representative of America.”

This proposition did some good, at least for the present moment. As very soon the delegates would not fail to fight with each other, and pull each other's hair where there was any to pull, it would be at the moment when it was necessary to elect a final buyer of this immovable region, so much disputed and so useless.

“In all cases,” cleverly remarked Dean Toodrink, “the United States will be entirely out of the question.”

“It seems to me very sensible,” said Eric Baldenak.

“Very handy,” said Col. Karkof.

“Right,” said Jan Harald.

“Mean,” said Jacques Jansen.

“Very English,” said Major Donellan.

Each one had given his opinion hoping to convince his colleagues.

“Then, gentlemen, it is perfectly understood that if we form a syndicate the rights of each State will be absolutely reserved for the future.” . . . It is understood. There was only to be found out what credit the different delegates had received from their governments. It was supposed that the whole when added up would represent such an enormous sum that there would not be the least doubt that the A.P.P.A. [N.P.P.A.] would fail to surpass this amount of money. This question of funds was met by Dean Toodrink.

Complete silence. Nobody would answer, show your pocketbook. Empty their pockets into the safes of a syndicate. Make known in advance how much each country would bid at the sale. No haste was shown. And if there should be a disagreement in this new-formed syndicate in the future, and circumstances should compel each one to make his own bids? And should the diplomat Karkof feel insulted at the trickery of Jacques Jansen, who would be insulted at the underhand intrigues of Jan Harald, who would refuse to support the high pretensions of Major Donellan, who, himself, would not stop to embroil each one of his associates. And now to show their credits – that was showing their play, when it was necessary to live up to it. There were really two ways only to answer the proper but indiscreet suggestion of Dean Toodrink. Either to exaggerate the credits, which would be very embarrassing, because it would then be the question of the payment, or to diminish them to such a point that they would be ridiculous and not to the purpose of the scheme. The ex-counsellor had this idea first, but it must be said to his credit, he did not seriously hold it. His colleagues, however, followed suit. "Gentlemen," said Holland, through its mouthpiece, "I regret, but for the purpose of the Arctic regions I can only dispose of fifty riechsthaler." "And I of thirty-five rubles," said Russia. "And I of twenty kronors," said Norway-Sweden. "And I of fifteen cronen," said Denmark. "Well," said Major Donellan, in a tone well befitting the disdainful attitude so common and natural to the English character, "then it would be better that you make the purchase, gentlemen, as England can only put up at the most one shilling and sixpence." And with this ironical remark the conference of the delegates of old Europe was at an end.

CHAPTER III

IN WHICH THE ARCTIC REGIONS ARE SOLD AT AUCTION *TO THE HIGHEST BIDDER.*

***

Why was this sale on the 3d of December going to be held in the regular auction hall, where usually only such objects as furniture, utensils, tools, instruments, etc., or art pieces, pictures, medals, and antique objects were sold? Why, so long as it was a piece of realty, was it not sold before a referee or a court of justice appointed for such sales? And why was the aid of a public auctioneer necessary when a part of our globe was going to be sold? How could this piece of the world be compared with any movable object when it was the most fixed thing on the face of the earth? Really, this seemed to be quite illogical, but it was not so, since the whole of the Arctic regions was to be sold in such a way that the contract would be valuable. Did this not indicate that in the opinion of the N.P.P.A. the immovable object in question contained something movable?

This singularity puzzled even the most eminently sagacious minds to be found in the United States. On the other hand, there had been such a sale in the past already. One portion of our planet had been sold in the auction-rooms with the help of a public appraiser to the highest public bidder. And this case had happened in America.

It was some years before, in San Francisco, Cal., an island of the Pacific Ocean, Spencer Island, was sold to the rich William W. Kolderup, who bid $500,000 in opposition to J.R. Taskinar, of Stockton. This island had cost $4,000,000. It is true it was an island which had inhabitants, only a few miles away from the coast, with forests, water, productive and solid, with fields and prairies, in condition to be cultivated, and not a vague wild region of water covered with eternal ice, defended by impenetrable icebergs, which very probably nobody would ever live in. It was thought, therefore, that the unknown polar region would never bring so high a price if sold at auction. Nevertheless, on the day of the sale a great many serious and curious people assembled to learn the end of the affair.

The sale could hardly fail to be interesting.

Ever since their arrival in Baltimore the European delegates had found themselves very much gazed at and always surrounded by many people. Of course they were much interviewed. In view of all this it was only natural that the public of America should have become very much excited. One manner of expressing the public excitement, very characteristic of Americans, was to make bets upon the result, an example which Europe quickly began to follow. Divided as the American citizens were into those of New England, those of the Middle States, Western States, and Eastern States, there was only one wish, and that was for the well-being of their country. But there was still a great deal of uncertainty. It was neither Russia, nor Sweden, nor Norway, nor Denmark, nor Holland, the chances of which they dreaded most. But it was England, with its territorial ambitions, with its only too well-known tendency to swallow everything, and its world-known Bank of England notes. Large sums of money were placed on the result. Bets were made on America and Great Britain, the same as on race-horses, and in almost all cases even money was put up. Some offers were made of 12 and 13[2] to 1 on Denmark, Sweden, Holland and Russia, but none would take such an offer.

The sale commenced at 12 o'clock.

Since early in the morning all business had been stopped in the street on account of the large crowd. By telegraph the papers were informed that most of the bets made by Americans had been taken up by the English, and Dean Toodrink immediately posted up a notice to that effect in the auction hall. The nearer the time came the higher grew the excitement. It was reported that the Government of Great Britain had placed large sums of money at the disposition of Major Donellan. "At the office of the Admiralty," observed one of the New York papers, "the Admirals pushed the sale as much as possible, as they hoped to figure conspicuously in the expeditions fitted out." How much truth there was in these stories no one knew. But the most conservative people in Baltimore thought that it was hardly possible that the amount of money at the command of the N.P.P.A. could cover the amount which would be bid by England, and therefore a very strong pressure was put on the Government of the United States at Washington to protect the interests of the society. In all this excitement the new society was represented by the single person, its agent, William S. Forster, who did not seem to worry at all over all these rumors and seemed quite confident of success.

As the time for the auction drew near the crowd grew larger. Three hours before the sale it was impossible to obtain admission to the auction hall. All the space set apart for the public was so much filled that there was danger that the building would fall in. There was only a small space left empty, surrounded by a railing, which had been reserved for the European delegates. They had just space enough to follow the progress of the sale, and were not even comfortably seated.

They were Eric Baldenak, Boris Karkof Jacques Jansen, Jan Harald, Major Donellan and his secretary, Dean Toodrink. They formed a solid group, standing together like soldiers on a battle-field. And were they not really going to battle for the possession of the North Pole? On the American side apparently nobody was represented. Only the codfish dealer was present and his face had an expression of the most supreme indifference.

He seemed little concerned and appeared to be thinking of his cargo which was to arrive by the next steamer. Where were the capitalists represented by this man, who, perhaps, was going to start millions of dollars rolling? This was such a mystery as to excite public curiosity to the utmost.

No one doubted that Mr. J.T. Maston and Mrs. Evangelina Scorbitt had something to do with the matter, but what could one guess on? Both were there, lost in the crowd, without any special place, surrounded by some members of the Baltimore Gun Club, friends of Mr. Maston. They seemed to be the least interested spectators in the hall. Mr. William S. Forster even did not seem to recognize them.

The auctioneer began by saying that contrary to the general rule it was impossible to show the article about to be sold. He could not pass from hand to hand the North Pole. Neither could they examine it nor look at it with a magnifying glass or touch it with their fingers to see whether the plating was real or only artificial, or whether it was an antique, which it really was, he said. It was as old as stone, it was as old as the world, since it dated back to the time the world was made.

If, however, the North Pole was not on the desk of the Public Appraiser, a large chart, clear in view of all interested persons, indicated with marked lines the parts which were going to be sold at auction. Seventeen degrees below the Polar Circle was a red line, clearly seen on the 84th parallel, which marked the section on the globe put up for sale. It appeared that there was only water in this region covered with ice of considerable thickness. But this was the risk of the purchaser. In any case he would not be disappointed in the nature of his merchandise by any misrepresentation.

At 12 o'clock exactly the public auctioneer entered by a little trap-door cut in the boards of the floor and took his place before the desk. His crier, Flint, had already arrived, and was walking up and down as agitated as a bear in a cage. Both were glad at the prospect, as they thought that the sale would run up to an enormous sum and put a large and acceptable percentage in their pockets. Of course the sale would have to be made under the regular, real American rule, "ready cash."

The amount of money, no matter how large it would be, must be raised by the delegates. At this moment a large bell ringing with vigor indicated that the bidding was going to begin. What a solemn moment! Many hearts quivered in that neighborhood. A minor riot spread among the crowd outside and reached into the hall, and Andrew R. Gilmour, the auctioneer, had to wait until quiet was restored. He got up and looked steadily at his assistants. Then he let his eyeglasses fall on his breast and said in as quiet a voice as possible: "Gentlemen, according to the plan of the Federal government, and thanks to the acquiescence given it by the European powers, we will sell a great fixed mass, situated around the North Pole, all that is within the limits of the 84th parallel, continents, waters, bays, islands, icebergs, solid parts or liquid, whatever they may be." Then, turning towards the wall, he said "Look at this chart, which has been outlined according to the latest discoveries. You will see that the surface of this lot contains 407,000 square miles. Therefore, to make the sale easier, it has been decided that the bids should be made for each square mile.

Each cent bid, for instance, will be equal to 407,000 cents and each dollar 407,000 dollars on the total purchase. A little silence, please, gentlemen."

This request was not superfluous, because the impatience of the public had reached such a degree that the voice of a bidder would hardly be heard. After partial silence had been established, thanks to the industry of the crier, Flint, who roared like a foghorn, Mr. Gilmour resumed as follows: "Before beginning I will mention only one condition of the sale. No matter what changes should happen, either from a geographical or meteorological standpoint, this region after being sold to the highest bidder is absolutely his own beyond any dispute, and the other countries have no right of possession whatever as long as the territory is not outside of the limits of the 84th degree north latitude." Again was this singular phrase mentioned at a very important moment. Some laughed at it, others considered it very seriously. "The bids are open," said the public auctioneer in a loud voice, and while his little ivory hammer was rolling in his hand he added in an undertone: "We have offers at 10 cents the square mile. Ten cents or the tenth part of a dollar – this would make an amount of $40,700 for the whole of this immovable property." Whether the auctioneer had had such offers or not does not matter, because the bid was covered by Eric Baldenak in the name of the Danish Government. "Twenty cents," said he.

"Thirty cents," said Jan Harald, for Sweden-Norway.

"Forty cents," said Col. Boris Karkof, for Russia.

This represented already a sum of $162,800 to begin with. The representative of England had not as yet opened his mouth, not even moved his lips, which were pressed tightly together. On the other side Wm. S. Forster kept an impenetrable dumbness. Even at this moment he seemed absorbed in reading a newspaper which contained the shipping arrivals and the financial reports of the markets each day.

"Forty cents per square mile," repeated Flint, in a voice which resembled a steam whistle, "40 cents."

The four colleagues of Major Donellan looked at each other. Had they already exhausted the credit allowed them at the beginning of the bidding? Were they already compelled to be silent?

"Go on, gentlemen," said the Auctioneer Gilmour, "40 cents. Who goes higher? Forty cents; why, the North Pole is worth much more than that, for it is guaranteed to be made of ice."

The Danish delegate said 50 cents and the Hollandish delegate promptly outbid him by 10 cents.

Nobody said a word. This 60 cents represented the respectable amount of $244,200. The lift given by Holland to the sale started a murmur of satisfaction. It seemed that the persons who had nothing in their pockets and nothing to their names were the most interested of all in this contest of dollars.

At the moment Jacques Jansen made his offer Major Donellan looked at his secretary, Dean Toodrink, and, with an almost imperceptible negative sign, kept him silent. Mr. William S. Forster, seeming very much interested in his paper, made some pencil notes. Mr. J.T. Maston, only replied to the smiles of Mrs. Evangelina Scorbitt with a nod of the head.

“Hurry up, gentlemen; a little life. Don't let us linger. This is very weak, very slow,” said the auctioneer. “Let me see. Nobody says more. Must I knock down the North Pole at such a price?” and as he spoke his hammer went up and down just like the cross in a priest's hands when he wishes to bless his people.

“Seventy cents,” said Jan Harald, with a voice which trembled a little.

“Eighty,” immediately responded Col. Boris Karkof.

“Hurry up, 80 cents,” said Flint, whose eyes were burning with excitement.

A gesture of Dean Toodrink made Major Donellan jump up like a spirit. “One hundred cents,” said he with a short and sharp tone, becoming in one who represented Great Britain. That one word made England responsible for $407,000.

The friends of the bidders for the United Kingdom made a great hurrah, which was repeated like an echo by the outside crowd. The bidders for America looked at each other with disappointment; $407,000; this was already a very large figure for such a region as the North Pole; $407,000 for ice, icebergs, and icefields?

And the man of the N.P.P.A. did not say one word, did not even raise his head. Would he decide to make at last one overwhelming bid? If he wanted to wait until the Danish delegates, those of Sweden, Holland, and Russia had exhausted their credit, it would seem that the proper moment had come. Their faces plainly showed that before the "100 cents" of Major Donellan, they had decided to quit the battlefield. "One hundred cents the square mile," said the auctioneer for the second time, "One hundred, one hundred, one hundred," cried out Flint, making a speaking-trumpet of his half-closed hand. "Nobody goes higher?" questioned Auctioneer Gilmour. "Is it heard? Is it understood? No regrets afterwards? We will sell it now." And he took his position and looking at his clerk, said: "once, twice"–

"One hundred and ten," very quietly said William S. Forster, without even raising his eyes, after having turned the page of his paper.

“Hip, hip, hip,” shouted the crowd who had put most of the money on America in the bets. Major Donellan was astonished. His long neck turned in all directions and he shrugged his shoulders, while his lips worked with great excitement. He tried to crush this American representative with one look, but without success, but Mr. Forster, cool as a cucumber, did not budge.

“One hundred and forty,” said Major Donellan.

“One hundred and sixty,” said Forster.

“One hundred and eighty,” said the Major, with great excitement.

“One hundred and ninety,” said Forster.

“One hundred and ninety-nine,” roared the delegate of Great Britain. With this he crossed his arms and seemed to defy the United States of America.

One might have heard a mouse run, or a miller fly, or a worm creep. All hearts were beating. A life seemed hanging on the lips of Major Donellan. His head, generally restless, was still now. Dean Toodrink had sat down and was pulling out his hairs one by one. Auctioneer Gilmour let a few moments run by. They seemed as long as centuries. The codfish merchant continued reading his paper and making pencil figures which had evidently nothing at all to do wth [with] the question. Was he also at the end of his credit? Did he intend to make a last offer? Did this amount of 199 cents for each square mile or $793,000 for the whole of the fixture at sale seem to him to have reached the last limit of absurdity?

“One hundred and ninety-nine cents,” repeated the public auctioneer. “We will sell it,” and his hammer fell on the table before him. “One hundred and ninety-nine,” cried the helper. “Sell it! Sell it!” And every one was looking at the representative of the N.P.P.A.

That surprising gentleman was blowing his nose on a large bandanna handkerchief, which nearly covered his whole face. Mr. J.T. Maston was looking at him intently, and so was Mrs. Evangelina Scorbitt. It could easily be seen by their anxious faces how much they tried to supress their violent emotion. Why was Forster hesitating to outbid Major Donellan? As for the imperturbable Forster, he blew his nose a second time, then a third time, with the noise of a real foghorn. But between the second and third blow he said as quietly as possible, with a modest and sweet voice. "Two hundred cents!"

A long shudder went through the hall. Then the American backers began to make such a noise that the very windows trembled. Major Donellan, overwhelmed, ruined, disappointed, had fallen into a seat by the side of Dean Toodrink, who himself was not in a much better condition. Two thousand miles at this price made the enormous sum of $814,000 and it was apparent that the credit of England did not permit them to overreach it.

"Two hundred cents," repeated the auctioneer. "Two hundred cents," said Flint. "Once, twice," said the auctioneer. "Does anybody go higher?" Major Donellan raised himself by a strong effort and looked wistfully at the other delegates. These only looked back their hope that by his bidding the Arctic region would escape the American bidder and would become the property of the European powers. But this was his last effort. The Major opened his mouth, closed it again, and in his person England sat down on itself. "Hip, hip, hurrah for the United States," roared the winners for victorious America. And in one instant the news of the purchase ran all over Baltimore, and by telegraph all over the United States, and by cable all over the Old World. Thus it was that the N.P.P.A. through the agency of William S. Forster, became the proprietor of the Arctic domain, including everything above the eighty-fourth parallel. The next day when Mr. Forster went to make his deposit for his purchase the name he gave was Impey Barbicane and the name of the house was Barbicane & Co.

CHAPTER IV

IN WHICH OLD ACQUAINTANCES APPEAR TO OUR NEW READERS, AND IN WHICH A WONDERFUL MAN IS DESCRIBED.

***

Barbicane & Co. The president of a gunning club. And really what had gunners to do in such an operation? You will see. Is it necessary to present formally Impey Barbicane, President of the Gun Club, of Baltimore, and Capt. Nicholl, and J.T. Maston, and Tom Hunter with the wooden legs, and the lively Bilsby, and Col. Bloomsberry, and the other associates? No, if these strange persons were twenty years older than at the time when the attention of the world was upon them they had always remained the same, always as much incomplete personally, but equally noisy, equally courageous, equally confusing when it was a question of some extraordinary adventure. Time did not make an impression on these gunners; it respected them as it respects cannons no longer in use, but which decorate museums and arsenals. If the Gun Club had 1,833 members in it when it was founded, names rather than persons, for most of them had lost an arm or leg, if 30,575 corresponding members were proud to owe allegiance to the Club, these figures had not decreased. On the contrary, and even thanks to the incredible attempt which they had made to establish direct communication between earth and moon, its celebrity had grown in an enormous proportion. No one can ever forget the report on this subject which was made by this Club and which deserves a few words of mention here.

A few years after the civil war certain members of the Gun Club, tired of their idleness, proposed to send a projectile to the moon by means of a Columbiad monster. A cannon 900 feet long, nine feet broad at the bore, had been especially made at Moon City and had then been charged with 400,000 pounds of gun-cotton.

From this cannon a small cylindro-conical bomb had been flung towards the stars with a pressure of six millards pounds per square inch. After having made a grand curve it fell back to the earth only to be swallowed up by the Pacific Ocean at

27° 7' of latitude and 41° 37' of longitude, west. It was in this region that the frigate, Susquehanna, of the American Navy, had fished it up to the surface of the ocean, to the great comfort of its occupants. Occupants? Yes, occupants; for two members of the Gun Club – its President, Impey Barbicane, and Capt. Nicholl – accompanied by a Frenchman well known for his boldness in such cases, had been in this flying-machine. All three of them came back well and healthy from this dangerous trip. But if the two Americans were here ready to risk any similar thing, the French Michel Ardan was not. On his return to Europe he brought a fortune with him, although it astonished a good many people, and now he is planting his own cabbage in his own garden, eating them and even digesting them, if one can believe the best-informed reporters.

After this discharge of the cannon, Impey Barbicane and Capt. Nicholl had lived on their reputation in comparative quietness. As they were always anxious to do another thing like it, they dreamt and tried to find out something else. Money they had in plenty. Out of five millions and a half which had been raised for them by subscription they had nearly $200,000 left. This money was raised in the Old and New Worlds alike. Besides, all they had to do was to exhibit themselves in their projectile in America and they could always realize large amounts of money. They had earned all the glory which the most ambitious mortal would look for. Impey Barbicane and Capt. Nicholl would have been well able to keep quiet and idle if this very idleness did not torment them. And it was simply no doubt to do something that they had gone to work and bought this part of the Arctic region.

But it must not be forgotten that if the purchase cost $800,000 and more, that it was Mrs. Evangelina Scorbitt who had put the necessary amount into this affair. Thanks to this generous woman Europe had been conquered by America. Since their return President Barbicane and Capt. Nicholl had enjoyed a supreme celebrity. But there was another man who deserved credit in the matter. It is easy to guess that J.T. Maston was the man of whom we speak, the temporary Secretary of the Gun Club. Did we not owe to this brilliant calculator all the mathematical formulae which enabled us to tell the story of the voyage to the moon? If he did not accompany his two associates on their terrible journey it was not fear which kept him back. No, indeed, it was only the injuries he had received during the war. For really it would have made a bad impression on the inhabitants on the moon to see him in his disabled condition as a representative of our people, and the moon only our humble satellite. To his extreme regret, Mr. Maston was compelled to stay at home. Nevertheless he had not been idle. After having constructed an immense telescope, which was put on the mountain of Long's Peak, one of the highest mountains of the Rocky range, he went up there personally, and after he had received the signal that the cannonball had been fired he did not once leave his post. From his place of observation he essayed the task of following his friends firing across the vast space. One might readily believe that his friends would be lost to the world; that it was very easily possible that this projectile could be compelled by the attraction of the moon to become a subsatellite. A deviation which one might call providential had changed the direction of the projectile, and after having made one trip around the moon, in place of touching it, it was carried away in a terrible fall and returned to us with a speed of 576,000 miles a minute to the moment when it was swallowed up by the ocean. Happy it was that the great liquid mass had deadened the fall, and that the American frigate Susquehanna was present at the fall. As soon as the news reached Mr. Maston, the Secretary of the Gun Club, he rushed with all possible haste from his observation point at Long's Peak to begin operations to save his friends. Divers were sent to the place where the projectile had fallen. And Mr. Maston even did not hesitate to put on a cork jacket to save and find his friends again. It was unnecessary to go to so much trouble. The projectile was found floating on the surface of the Pacific Ocean after having made its beautiful fall. And President Barbicane with Capt. Nicholl and Michel Ardan were found playing dominoes in their floating prison on the surface of the ocean.

To return to J.T. Maston, it is proper to say that his part in the affair deserves a good deal of credit. It is certain that he was not pretty with his metallic arm. He was not young, fifty-eight years old, at the time we write this story. But the originality of his character, the vivacity of his intelligence, the vigor which animated him, the ardor which he had in all such things, had made him the ideal of Mrs. Evangeline Scorbitt. His brain carefully hidden under his cover of gutta-percha was yet untouched, and he would still pass as one of the most remarkable calculators of his age.

Mrs. Evangelina Scorbitt, although the least figuring gave her a headache, had yet a great liking for mathematicians, even if she had no taste for mathematics. She considered them a higher and more endowed race of human beings. Heads where the X, Y, Z were mixed up like nuts in a barrel, brains which played with signs of algebra, hands which juggled with the integral triples, these were what she liked.

Yes, these wise people seemed to her worthy of all admiration and support. She felt herself drawn strongly towards them. And J.T. Maston was exactly that kind of man and one she adored, and her happiness would be complete if they two could be made one. This was the end of her mathematics. This did not disturb the Secretary of the Gun Club, who had never found happiness in unions of this kind.

Mrs Evangelina Scorbitt was not young any more. She was forty-five years old, had her hair pasted on her temples, like something which had been dyed and re-dyed; she had a mouth full of very long teeth, with not one missing; her waist was without shape, her walk without grace; in short, she had the appearance of an old maid, although she had been married only a few years before she became a widow. She was an excellent person withal, and nothing was lacking in her cup of happiness except one thing, namely, that she wished to make her appearance in the society of Baltimore as Mrs. J.T. Maston. Her fortune was very considerable. She was not rich like the Goulds, Mackays, or Vanderbilts, whose fortunes run into the millions, and who might give alms to the Rothschilds. Neither did she possess three hundred millions like Mrs. Stewart, eighty millions like Mrs. Crocker, and two hundred millions like Mrs. Carper. Neither was she rich like Mrs. Hamersley, Mrs. Hetty Green, Mrs. Mafitt, Mrs. Marshall, Mrs. Paran Stevens, Mrs. Minturn, and many others. At any rate she had a right to take a place at that memorable feast at the Fifth Avenue Hotel in New York, where there were only admitted as guests people who had at least five millions. In brief, Mrs. Evangelina Scorbitt had four millions of good sound dollars, or twenty millions of francs, which came to her from John P. Scorbitt, who made his fortune both in the business of selling dry goods and salt pork. Well, this fortune this generous widow would have been glad to use for the profit of J.T. Maston, to whom she would also bring a treasure of tenderness much more inexhaustible.

Therefore when Mrs. Evangelina Scorbitt had heard of the requirements of Mr. Maston she had gladly agreed to put a few hundred thousand dollars in the affair of the N.P.P.A. without having the least idea of what they intended to do with it. It is true she was convinced that as long as J.T. Maston had something to do with the affair it could not help being grand, sublime, superhuman, etc. Thinking of the Secretary was for her future enough. One might think that after the auction sale, when it was declared that Barbicane & Company would be the name of the new firm, and it would be presided over by the President of the Gun Club, she would enjoy Mr. Maston's whole confidence. Was she not at the same time the largest stockholder in the affair? So it came about that Mrs. Evangelina Scorbitt found herself proprietress of this polar region, all beyond the line of the eighty-fourth parallel. But what would she do with it? Or rather, what profit would the Society get out of it? This was the question; and if it interested Mrs. Evangelina Scorbitt from a financial standpoint it interested the whole world, more on account of the general curiosity about the whole mystery. This excellent woman, otherwise very discreet, had often tried to get some information from Mr. Maston on this subject before putting money at the disposal of the purchasers. Without a doubt there was a grand enterprise, which, as Jean Jacques said, has never had its like before, and would never have any imitation, and which would far outshine the reputation made by the Gun Club in sending a projectile to the moon and trying to get in direct communication with our satellite. But when she persists with her queries Mr. Maston invariably replied: “Dear madame, have patience,” And if Mrs. Evangelina Scorbitt had confidence before, what an immense joy did she feel when the triumph which the United States of America had won over the combined European powers. “But shall I not finally know the object?” asked she, smiling at the eminent calculator.

“You will soon know it,” answered Mr. Maston, shaking heartily the hand of his partner – the American lady.

This calmed for the moment the impatience of Mrs. Evangelina Scorbitt. A few days afterwards the Old and New World were shaken up quite enough when the secret object of the company was announced, and for the realization of which the N.P.P.A. made an appeal to the public for a subscription.

The Society had purchased this portion of the circumpolar region to make use of the coal mines of the North Pole.

CHAPTER V

IN WHICH THE POSSIBILITY THAT COAL MINES SURROUND *THE NORTH POLE IS CONSIDERED.*

***

Are there coal mines at the North Pole? This was the first question suggested to intelligent people. Some asked why should there be coal mines at the North Pole? Others with equal propriety asked why should there not be? It is well known that coal mines are spread all over the world. There are many in different parts of Europe. America also possesses a great many, and it is probable that the United States mines are the richest of all. There are also many in Asia, Africa, and Australia. The more our globe becomes known the more mines are discovered. We will not be in need of coal for at least hundreds of years to come. England alone produces 160,000,000 tons every year, and over the whole world it is estimated 400,000,000 tons are yearly produced. Naturally, this coal output must grow every year in proportion with the constantly increasing industries. Even if electricity takes the place of steam, it will still be necessary to use coal. We are so much in need of it that the world might be called "an animal of coal," and therefore it is necessary to take good care of it. Coal is used not only as a fuel, but also as a crude substance of which science makes great use. With the transformations to which it has been submitted in the laboratory, it is possible to paint with it, perfume with it, purify, heat, light with it, and even beautify a diamond with it. It is as useful as iron or even more so. It is fortunate that this last-mentioned metal will never be exhausted, as really the world is composed of it. The world might be considered a vast mass of iron, as other metals, and even water and stone, stand far behind it in the composition of our sphere. But if we are sure of a continuous supply for our consumption of iron, we are not so of coal. Far from it. People who are competent to speak, and who look into the future for hundreds of years, always allude to this coal famine. "But," say the opposing party – and in the United States there are many people who like to contradict for the mere sake of argument, and who take pleasure in contradicting – "Why should there be coal around the North Pole?"

"Why?" answered those who took the part of President Barbicane, "because, very probably at the geological formation of the world, the sun was such that the difference of temperature around the equator and the poles were not appreciable. Then immense forests covered this unknown polar region a long time before mankind appeared, and when our planet was submitted to the incessant action of heat and humidity. This theory the journals, magazines, and reviews publish in a thousand different articles either in a joking or serious way. And these large forests, which disappeared with the gigantic changes of the earth before it had taken its present form, must certainly have changed and transformed under the lapse of time and the action of internal heat and water into coal mines. Therefore nothing seemed more admissible than this theory, and that the North Pole would open a large field to those who were able to mine it. These are facts, undeniable facts. Even people who only calculated on simple probabilities could not deny them. And these facts led many people to have great faith in them.

It was on this subject that Major Donellan and his secretary were talking together one day in the most obscure corner of the "Two Friends" inn.

"Well," said Dean Toodrink, "there is a possibility that this Barbicane (who I hope may be hanged some day) is right."

"It is probable," said Major Donellan, "and I will almost admit that it is certain. There will be fortunes made in exploring this region around the pole. If North America possesses so many coal mines and, according to the papers, new ones are discovered quite frequently, it is not at all improbable that there are many yet to be discovered. I may add that Prof. Nordenskiol has found many kinds of different stones which contain a great variety of fossil plants in his researches in the Arctic region."

"Higher up?" asked Dean Toodrink.

"Higher up, or rather further up, in a northerly direction," answered the Major, "the presence of coal is practically established, and it seems as if you would only have to bend down to pick it up. Well, if coal is so plentiful on the surface of these countries, it is right to conclude that its beds must go all through the crust of the globe." He was right. Major Donellan knew the geological formation around the North Pole well, and he was not a safe person to dispute this question with. And he might have talked about it at length if other people in the inn had not listened. But he thought it better to keep quiet after asking: "Are you not surprised at one thing? One would expect to see engineers or at least navigators figure in this matter, while there are only gunners at the head of it?"

It is not surprising that the newspapers of the civilized world soon began to discuss the question of coal discoveries at the North Pole.

"And why not," asked the editor of an American paper who took the part of President Barbicane, "when it is remembered that Capt. Nares, in 1875 and 1876, at the eighty-second degree of latitude, discovered large flower-beds, hazel trees, poplars, beech trees, etc.?"

"And in 1881 and 1884," added a scientific publication of New York, "during the expedition of Lieut. Greely at Lady Franklin Bay, was not a layer of coal discovered by our explorers a little way from Fort Conger, near Waterhouse? And did not Dr. Pavy say that these countries are certainly full of coal, perhaps placed there to combat at some day the terrible masses of ice which are found there?”

Against such well-established facts brought out by American discoverers the enemies of President Barbicane did not know what to answer. And the people who asked why should there be coal mines began to surrender to the people who asked why should there be none. Certainly there were some, and very considerable ones, too. The circumpolar ice-cap conceals precious masses of coal contained in those regions where vegetation was formerly luxuriant. But if they could no longer dispute that there were really coal mines in this Arctic region the enemies of the association tried to get revenge in another way. “Well,” said Major Donellan one day after a hard discussion which had arisen in the meeting-room of the Gun Club and during which he met President Barbicane face to face, “all right. I admit that there are coal mines; I even affirm it, there are mines in the region purchased by your society, but go and explore them – ha! ha! ha!”

“That is what we are going to do,” said Impey Barbicane.

“Go over the eighty-fourth degree, beyond which no explorer as yet has been able to put his foot?”

“We will pass it – reach even the North Pole,” said he. “We will reach it.” And after hearing the President of the Gun Club answer with so much coolness, with so much assurance, to see his opinion so strongly, so perfectly affirmed, even the strongest opponent began to hesitate. They seemed to be in the presence of a man who had lost none of his old-time qualities, quiet, cold, and of an eminently serious mind, exact as a clock, adventurous, but carrying his practical ideas into the rashest enterprises.

Major Donellan had an ardent wish to strangle his adversary. But President Barbicane was stout and well able to stand against wind and tide, and therefore not afraid of the Major. His enemies, his friends and people who envied him knew it only too well. But there were many jealous people, and many jokes and funny stories went round in regard to the members of the Gun Club. Pictures and caricatures were made in Europe and particularly in England, where people could not get over the loss which they suffered in the matter of pounds sterling. “Ah,” said they, “this Yankee has got it in his head to reach the North Pole. He wants to put his foot where, up to the present time, no living soul has yet been. He wants to build palaces and houses and, perhaps, the White House of the United States, in a part of the world which has never yet been reached, while every other part of the world is so well known to us." And then wild caricatures appeared in the different newspapers. In the large show-windows and news-depots, as well in small cities of Europe as in the large cities of America, there appeared drawings and cartoons showing President Barbicane in the funniest of positions trying to reach the North Pole. One audacious American cut had all the members of the Gun Club trying to make an underground tunnel beneath the terrible mass of immovable icebergs, to the eighty-fourth degree of northern latitude, each with an axe in his hand. In another, Impey Barbicane, accompanied by Mr. J.T. Maston and Capt. Nicholl, had descended from a balloon on the much-desired point, and after many unsuccessful attempts and at the peril of their lives, had captured a piece of coal weighing about half a pound. This fragment was all they discovered of the anticipated coal-fields. There were also pictures made of J.T. Maston, who was as much used for such purposes as his chief. After having tried to find the electric attraction of the North Pole, the secretary of the Gun Club became fixed to the ground by his metallic hand.

The celebrated calculator was too quick-tempered to find any pleasure in the drawings which referred to his personal conformation. He was exceedingly annoyed by them, and Mrs. Evangelina Scorbitt, it may be easily understood, was not slow to share his indignation. Another drawing in the Lanterne of Brussels represented the members of the Council and the members of the Gun Club tending a large number of fires. The idea was to melt the large quantities of ice by putting a whole sea of alcohol on them, which would convert the polar basin into a large quantity of punch. But of all these caricatures, that which had the largest success was that which was published by the French *Charivari*, under the signature of its designer, "Stop." In the stomach of a whale Impey Barbicane and J.T. Maston were seated playing checkers and waiting their arrival at a good point. The new Jonah and his Secretary had got themselves swallowed by an immense fish, and it was in this way, after having gone under the icebergs, that they hoped to gain access to the North Pole. The President of this new Society did not care much about these pictures, and let them say and write and sing whatever they liked.

Immediately after the concession was made and the Society was absolute master of the northern region, appeal was made for a public subscription of $15,000,000. Shares were issued at $100, to be paid for at once, and the credit of Barbicane & Co. was such that the money ran in as fast as possible. The most of it came from the various States of the Union. “So much the better,” said the people on the part of the N.P.P.A. “The undertaking will be entirely American.”

So strong, indeed, were the foundations upon which Barbicane & Co. were established that the amount necessary to be subscribed was raised in a very short time, and even thrice the amount. Everybody was interested in the matter, and the most scientific experts did not doubt its success.

The shares were reduced one-third, and on Dec. 16 the capital of the Society was $15,000,000 in cash. This was about three times as much as the amount subscribed to the credit of the Gun Club when it was going to send a projectile from the earth to the moon.

CHAPTER VI

IN WHICH A TELEPHONE COMMUNICATION BETWEEN *MRS. SCORBITT AND J.T. MASTON IS INTERRUPTED.*

***

President Barbicane was not only convinced that he would reach his object when the amount which had been raised took another obstacle out of his way. Had he not been perfectly sure of success he would not have made an application for a public subscription. And now the time had come when the North Pole would be conquered. It was felt certain that President Barbicane and his Council of Administration had means to succeed where so many others had failed. They would do what neither Franklin, nor Kane, nor De Long, nor Nares, nor Greely had been able to accomplish. They would pass the 84th parallel, they would take possession of the vast region purchased at an auction sale, they would make this country the thirty-ninth star in the flag of the American Union.

“Fake,” was all that the European delegates and their friends in the Old World could say. Nothing was more true, however, and this practical, logical means of conquering the North Pole, which was so simple that it was almost childish, was one which J.T. Maston had suggested to them. It was that brain, where ideas were constantly evolving, which had laid out this great geographical project in a way which could not but succeed.

It cannot be too often repeated that the Secretary of the Gun Club was a remarkable calculator, we might say a postgraduate calculator. But a single day was needed by him to solve the most complicated problems in mathematical science. He laughed at these difficulties whether in algebra or in plain mathematics. You should have seen him handle his figures, the signs which make up algebra, the letters in the alphabet, representing the unknown quantities, the square or crossed lines representing the way in which quantities are to be operated. All signs and lines, and radicals used in this complex language were perfectly familiar to him. And how they flew around under his pen, or rather under the piece of chalk which he attached to his iron hand, because he preferred to work on a blackboard. And this blackboard, six feet square, this was all he wanted, he was perfectly at home in his work. Nor was it figures alone which he used in his calculations. His figures were fantastic, gigantic, written with a practiced hand. His "2" and "3" were as nice and round as they could be, his 7 looked like a crutch and almost invited a person to hang on it. His 8 was as well formed as a pair of eye-glasses; and the letters with which he established his formulas, the first of the alphabet, a, b, c, which he used to represent given or known quantities, and the last, x, y, z, which he used for unknown quantities to be discovered, particularly the "z," and those Greek letters

δ, ω, α.

Really an Archimedes might have been well proud of them. And those other signs, made with a clean hand and without fault, it was simply astonishing. His + showed well that this sign meant an addition of one object to the other, his –, if it was a little smaller, was also in good shape. His =, too, showed that Mr. Maston lived in a country where equality was not a vain expression, at least amongst the people of the white race. Just as well were his > and his < and his ::, used in expressing proportions. And the √ , which indicated the root of a certain number or quantity, it was to him a mark of triumph, and when he completed it with a horizontal line in this √‒, it seemed as if this outline on his blackboard would compel the whole world to submit to his figuring.

But do not think that Mr. J.T. Maston's mathematical intelligence was confined to elementary algebra! No; no matter what figuring he had to do, it was alike familiar to him, and with a practised hand he made all the signs and figures, and even did not hesitate at ∫ which looks very simple, but behind which lays a great deal of calculation. The same with the sign ∑, which represents the sum of a finished number. Also the sign ∞, by which the mathematicians designate the incomplete, and all those mysterious symbols which are used in this language and which are unknown to the common people. This astonishing man was able to do anything even in the very highest grades of mathematics. Such was J.T. Maston. And therefore it was that his associates had such perfect confidence in him when he set out to figure the most difficult problems in his audacious brain. This it was which led the Gun Club to trust him with the difficult problem of sending a projectile to the moon. And this was why Mrs. Evangelina Scorbitt, jealous of his fame, felt for him an admiration which ended in love. In this present case – that is, how to solve the conquering of the North Pole, J.T. Maston had but to begin to think and dream himself into the Arctic regions. To reach the solution the secretary had but to undertake certain mathematical problems, very complicated, perhaps, but over which in all cases he would come out ahead.

It was safe to trust Mr. J.T. Maston, even where the smallest and simplest mistake would have meant a loss of millions. Never, since the time his youthful brain began to think of mathematics had he committed a mistake – not even one of a thousandth of an inch – if his calculations were made up on the length of an object. If he had made a mistake of only the smallest amount he would have torn his gutta-percha cap from his head. Now let us see him while engaged in his calculations, and for this purpose we must go back a few weeks.

It was about a month before the publication of the circular addressed to the inhabitants of the Old and New Worlds that Mr. Maston had undertaken to figure out the elements of a scheme in which he had promised his associates the greatest success. For a number of years Mr. Maston had lived at No. 179 Franklin Street, one of the most quiet streets of Baltimore, far away from the centre of business, for which he did not care anything, far away from the noise of the great crowd, which disgusted him. There he occupied a modest little house known by the name of "Ballistic Cottage," having for his income only his pension allowed to him as a retired officer of artillery and the salary which he received as Secretary of the Gun Club. He lived alone, served by his Negro "Fire-Fire." This Negro was not an ordinary servant; he was rather an appreciative friend and treated his master as if he were his own brother. Mr. Maston was a decided bachelor, having an idea that being a bachelor was the only sensible way of living in the present world. He knew the proverb, "a woman can draw more with one hair than four oxen at the plough," and he disproved it. If he occupied his cottage alone it was only because he wished to do so. We know that he only had to make the motion to change his solitude of one to a company of two and his small income to the income of a millionaire. He did not doubt it. Mrs. Evangelina Scorbitt would have been only too happy to . . . But up to this time Mr. Maston had not been happy to . . .

and it seemed certain that these two beings, so well made one for the other, at least this was the opinion of the tender widow, would never reach the transformation period. The cottage was a very simple one. A ground-floor, with a veranda and a floor over it; a small parlor and small dining-room, with a kitchen and another room in an outbuilding stand at the back of the garden. Upstairs his sleeping-room and his working-studio, looking on the garden, and where the noise of the outer world could never penetrate. Within these walls there had been made calculations which would have made Newton, Euclid, or Laplace jealous. How different was the mansion of Mrs. Evangelina Scorbitt, situated in the richest quarter of New Park, with facade of balconies, finished in the finest architecture, half Romanesque, half Gothic, with its richly-furnished apartments, its grand halls, its picture galleries, in which French artists held the highest places; its magnificent staircase, its great number of domestics, its stables, its coach-houses, its garden, with the finest of flowers, statues, fountains, and the tower on the top of the building, on which the blue and gold coat-of-arms of the Scorbitt family was upon a glittering banner. Three miles, three long miles at least separated the Palace at New Park from the "Ballistic Cottage." But a private telephone wire connected there, and in answer to "Hello! hello!" a conversation could be carried on between the mansion and cottage. If the persons could not look at each other they could at least hear each other. It will astonish none to hear that time upon time Mrs. Scorbitt began talking and ringing on the telephone to Mr. Maston when he was busily engaged with his figures. Then the calculator had to quit his work with some reluctance. He received a friendly "How do you do?" from Mrs. Scorbitt, which he answered with a grunt, which was sweetened into a kindly greeting by the distance over the telephone. After a conversation he was glad to go back to his figures. It was on the 3d of December, after a long and last conference, that Mr. Maston took leave of his friends and members of the Club to begin to do his share of the work. It was a very important work with which he had charged himself, for it was the question of figure mechanical appliance which would enable him to gain access to the North Pole, and which would allow him to make use of those large fields of snow now covered with impenetrable ice. He estimated that he needed at least a week to accomplish this mysterious calculation, exceedingly complicated and delicate to handle, involving several deep and important problems. Therefore, to avoid all unnecessary annoyance, it had been decided that the Secretary of the Gun Club should retire to his cottage and that he should not be disturbed by any one. This was a great disappointment for Mrs. Evangelina Scorbitt, but she was compelled to accept it. While President Barbicane, Capt. Nicholl and his associates, the jolly Bilsby, Col. Bloomsberry, Tom Hunter, with the wooden legs, were all saying their good-bys to him and wishing him success, Mrs. Evangelina Scorbitt appeared and made her last visit to Mr. Maston.

“You will succeed, my dear,” said she at the moment of separation.

“And above all do not make a mistake,” added President Barbicane. “A mistake? He?” exclaimed Mrs. Scorbitt.

“No more than God has made a mistake in putting together this world,” modestly answered the Secretary. Then, after shaking hands all around and after several more sighs and wishes of success and suggestions not to make too severe a work of it, the calculator was left alone. The door of the Ballistic cottage was closed and Fire-Fire had orders to admit none, not even if the President of the United States should ask admission.

During his first two days of seclusion J.T. Maston thought and thought, without even touching the piece of chalk, upon the problem which he had taken on himself. He consulted certain books relative to the elements, the earth – its size, its thickness, its volume, its form, its rotation upon its axis – all elements which he had to use as the basis of his calculations.

The principles of these elements which he used, and which we put before the reader, were as follows:

> Form of the earth: An ellipsis of revolution the longest radius of which is 6,377,398 metres; the shortest, 6,356,080 meters. The circumference of the earth at the equator, 40,000 kilometres. Surface of the earth, approximate estimate, 510,000,000 of square kilometers. Bulk of the earth, about 1,000 millards of cubic kilometres; that is, a cube having a metre in length, height, and thickness. Density of the earth, about five times that of the water. Time of the earth on the orbit around the sun, 365

days, 6 hours, 9 minutes, 10 seconds, 37 centimes. This gives the globe a speed of 30,400 miles travelled over by the rotation of the earth upon its axis. For a point of its surface situated at the equator, 463 meters per second. These were the principal measures of space, time, bulk, etc., which Mr. Maston used in his calculations.

It was the 5th of October, about 5 o'clock in the afternoon, it is important to mention, when this remarkable work was begun, when J.T. Maston began to work upon it. He began his calculation with a diagram representing the circumference of the earth around one of its grand circles, say the equator. The blackboard was there, in a corner of his study, upon a polished oak easel, with good light shining on it, coming by one of the windows near by. Small pieces of chalk were on the board attached to the stand. The sponge was near the hand of the calculator. His right hand, or rather his right hook, was all ready for the placing of figures which he was going to use. Standing up, Mr. Maston made a large round circle, which represented the world. The equator he marked by a straight line. Then in the right corner of the blackboard he began to put the figures which represented the circumference of the earth:

40,000,000.

This done, he began figuring on his problem. He was so much occupied by it that he had not observed the weather without. For an hour a storm had raved through the country which affected all living beings. It was a terrific storm, the rain was falling in torrents, everything seemed turned upside down in nature. Two or three times lightning had illuminated the scene around him. But the mathematician, more and more absorbed in his work, saw and heard nothing. Suddenly an electric bolt, attracted by the lightning outside, sparkled in his room, and this disturbed the calculator. "Well," said Mr. Maston, "if unwelcome visitors cannot get in by the door they come by telephone. A nice invention for people who wish to be left alone. I will go to work and cut off the electric wire, so I will not be disturbed again while my figuring lasts." With this he went to the telephone and said sternly: "Who wants to talk to me? Just make it short." The reply came back: "Did you not recognize my voice, my dear Mr. Maston? It is I,

Mrs. Scorbitt." "Mrs. Scorbitt! She will never give me a moment's rest," uttered Mr. Maston to himself in a low voice that she could not hear. Then he thought he should at least answer her in a polite manner, and said: "Oh, is that you, Mrs. Scorbitt?"

"Yes, dear Mr. Maston."

"And what can I do for Mrs. Scorbitt?" asked Maston.

"I want to tell you that a terrible storm and lightning is destroying a large part of our city." "Well," he replied, "I cannot help it." "But I want to ask whether you have thought to close your windows?" Mrs. Scorbitt had hardly finished her sentence when a terrible thunderbolt struck the town. It struck in the neighborhood of the Ballistic cottage, and the electricity, passing along the wire with which the telephone was provided, threw the calculator to the floor with a terrible force. J.T. Maston made the best summersault he ever did in his life. His metal hook had touched the live wire and he was thrown down like a shuttlecock. The blackboard, which he had struck in his fall, was sent flying to another part of the room. Then the electricity passed into other objects and disappeared through the floor. The stupefied Mr. Maston got up and touched the different parts of his body to assure himself that he was not hurt internally. This done, he resumed his cold, calculating way. He picked everything up in his room, put it in the same place where it had been before and put his blackboard on the easel, picked up the small pieces of chalk and began again his work, which had been so suddenly interrupted. He noticed that on account of the fall the number which he had made on the right side of the blackboard was partly erased, and he was just about to replace it when his telephone again rang with a loud noise. “Again,” said J.T, Maston, and going to the telephone he exclaimed, “who is there?” “Mistress Scorbitt.” “And what does Mrs. Scorbitt want?” “Did not this terrible thunderbolt strike Ballistic cottage? I have good reason to think so. Ah, great God, the thunderbolt!”

“Don't be alarmed, Mrs. Scorbitt.”

“You have not been injured, Mr. Maston?”

“Not at all,” he replied.

“You are sure you have no injuries whatever,” said the lady.

“I am only touched by your kindness towards me,” replied Mr. Maston, thinking it the best way to answer.

“Good evening, dear Mr. Maston.”

“Good evening, dear Mrs. Scorbitt.”

Returning to his work Mr. Maston said, sotto voce, “To the devil with her. If she had not handled the telephone at such a time I would not have run the risk of being hurt by electricity.”

Mr. Maston did not wish to be interrupted in his work again and so took down his telephone and cut the wire. Then, taking again as basis the figure which he had written, he added different formulas of it, and finally a certain formula which he had written on his left side, and then he began to figure in all the language of algebra. A week later, on the 11th of October, this magnificent calculation was finished and the Secretary of the Gun Club brought his solution of the problem with great pride and satisfaction to the members of the Gun

Club, who were awaiting it with very natural impatience. This then was the practical way to get to the North Pole mathematically discovered. Here was also a society, under the name of the N.P.P.A., to which the Government of Washington had accorded a clear title of the Arctic region in case they should buy it on auction, and we have told of the purchase made in favor of American buyers and of the appeal for a subscription of $15,000,000.

CHAPTER VII

IN WHICH PRESIDENT BARBICANE SAYS NO MORE THAN SUITS HIS PURPOSE.

***

On the 22nd of December the subscribers to Barbicane & Co. were summoned to a general meeting. It is hardly necessary to say that the headquarters of the Gun Club were selected as the place of the meeting. In reality the whole block would not have been sufficient to give room to the large crowd of subscribers who assembled on that day. But a meeting in the fresh air on one of the public squares of Baltimore was not very agreeable in such cold weather. Usually the large hall of the Gun Club was decorated with models of all kinds lent by members of the Club. It was a real museum of artillery. Even the furniture, chairs and tables, sofas and divans, recalled by their strange shapes those murderous engines which had sent into a better world many brave people whose greatest wish was to die of old age.

On this meeting day all these things were taken down and out. This was not a meeting for the purpose of war, but a commercial and peaceful meeting over which Impey Barbicane was going to preside. All room possible had been made for the subscribers who arrived from all parts of the United States. In the hall as well as in the adjoining rooms the crowds were pushing and pressing each other without heeding the innumerable people who were standing on the adjoining streets. The members of the Gun Club, as first subscribers to the affair, had places reserved for them very near the desk. Among them could be found Col. Bloomsberry, more happy than ever; Tom Hunter, with his wooden legs, and the jolly Bilsby. Very snug in a comfortable armchair was Mrs. Evangelina Scorbitt, who should really have had a place on the right hand of the President, as she was in reality the owner of the Arctic region. Several other ladies were in the crowd. They could readily be seen by their large and much-decorated hats in many different colors. The large crowd on the outside tried to push into the hall, and one might easily have thought that all the people present were not merely helpmates of the members of the Gun Club, but rather their personal friends. The European delegates – Swedish, Norwegians, Danish, English, Dutch, and Russian – occupied reserved seats, and if they had bought any stock in this society it was only each one individually to such an extent as to justify a vote in the proceedings. After they had been so closely united in purchasing these regions they were united now only to annoy the purchasers. It may easily be imagined what intense curiosity they had to hear the important communication which the President was about to make to them. This communication undoubtedly would throw some information on the point as to how the society would proceed to reach the North Pole. Was this not a more difficult thing than merely to make use of the coal mines? If there should be any objections to make you may be assured that Major Donellan, backed up by his secretary, Dean Toodrink, would make them, and the other delegates would not be very slow in adding their word also. The Major had firmly decided to harass and annoy his rival, Impey Barbicane, as much as he possibly could.

It was 8 o'clock in the evening. The hall, the parlors, and all quarters occupied by the Gun Club blazed with lights which the Edison electroliers throw out. As soon as the doors were thrown open for the public a terrible crowd jammed into the hall. But everyone became silent as the ushers announced that the Council of Administration was coming. There, on a draped platform, with a table covered with black cloth, in full sight, President Barbicane, his Secretary, J.T. Maston, and his associates, took their places. A triple round of cheers, followed by hearty "tigers," rang through the hall and out to the adjoining streets. Very solemnly Mr. Maston and Capt. Nicholl took their seats. Then the President, who had remained standing, opened the proceedings. He put his right hand in his trouser's pocket and his left hand in his vest front and began as follows:

"Lady and gentlemen subscribers, the Council of Administration has called a meeting in these headquarters of the Gun Club to make an important communication to you. You have learned by the circulars and through the discussions in the papers that the object of our Club is to explore the large coal fields situated in the Arctic regions, which we have recently purchased and to which we hold a title from the American Government. The amount of money raised by public subscription will be used for these purposes. The success which will be attained by it surpasses belief and the dividends your money will bring you will be unsurpassed in the commercial or financial history of this or any other country." Here applause was heard for the first time and for a moment the orator was interrupted. "You do not forget," said he, "how we have proved to you that there must be vast coal fields in these regions, perhaps also fields of fossil ivory. The articles published on this subject do not allow any doubt that coal fields are there, and coal is now, you know, the basis of all our commercial industry. Without mentioning the coal which is used every year in firing and heating, we might think of coal used for many other purposes, of which I could mention a hundred different ones. It is certain that coal is the most precious substance, and will some day, on account of the large consumption of it; fail in its supply. Before 500 years have passed the coal mines which are at present in use will have stopped giving coal."

"Three hundred years," cried one of those present. "Two hundred years," answered another.

"Let us say at some time sooner or later," continued the President, calmly, "and let us suppose, too, that we will even discover new coal fields yet, whose coal will give out, say at the end of the nineteenth century." Here he stopped to give his listeners a chance to grasp the idea. Then he began again: "Therefore, we come here, subscribers, and I ask you to rise and go with me to the North Pole immediately." Everybody present got up and seemed about to rush away and pack their trunks, as if President Barbicane had a vessel ready to take them direct to the North Pole. But a remark made by Major Donellan in a clear and loud voice brought them back to reality and stopped them at once. "Before starting" he asked, "I would like to know by what means we can reach the North Pole?"

"Either by water, or land, or by air," quietly answered President Barbicane.

All the people present sat down, and it may readily be understood with what a feeling of curiosity.

"In spite of all the devotion and courage of previous explorers, the eighty-fourth parallel has thus far been the northern limit reached. And it may fairly be supposed that this is as far north as anybody will ever get by the means employed at the present day. Up to the present time we have only used boats and vessels to reach the icebergs, and rafts to pass over the fields of ice. People should not adopt such rash means and face the dangers to which they are exposed through the low temperature. We must employ other means to reach the North Pole."

It could be seen by the excitement which took hold of the auditors, that they were on the point of hearing the secret which has been so vigorously searched for by every one.

"And how will you reach it?" demanded the delegate of England.

"Before ten minutes have passed you will know it, Major Donellan," said President Barbicane, "and I may add in addressing myself to all the stockholders, that they should have confidence in us as the promoters of this affair, for we are the same who have tried to send a projectile to the moon."

"Yes," cried Dean Toodrink, sarcastically, "they tried to go as far as the moon. And we can easily see that they are here yet."

President Barbicane ignored the interruption. Shrugging his shoulders, he said in a loud voice:

“Yes, ladies and gentlemen, in ten minutes you will know what we are going to do.”

A murmur, made up of many “Ahs!” and “Ohs!” followed this remark. It seemed to them as if the orator had said in ten minutes they would be at the North Pole. He then continued in the following words:

“First of all, it is a continent which forms this arctic region, or it is an ocean, and has Commander Nares been right in calling it 'paleocrystic ocean,' which means an ocean of old ice? To this question I must answer that I think he was not right.”

This is not sufficient,” exclaimed Eric Baldenak. “It is not the question of supposing, it is the question of being certain.”

“Well, we are certain,” came the answer to this furious inquirer. “Yes, it is a solid continent and not an ice ocean which the N.P.P.A. has purchased and which now belongs to the United States and which no European power has the right to touch.”

A little murmur came from the neighborhood of the delegates of the Old World. “Bah!” they said.

“It is full of water, a regular washbasin which you will not be able to empty.” Dean Toodrink as usual made most of the remarks and met the hearty applause of his associates. “No, sir,” answered President Barbicane, quickly. “There is a regular continent, a platform which rises like the Gobi desert in Central Asia, three or four kilometres above the surface of the ocean. This is very easy to be seen from the observations made in the neighboring countries, of which the polar region is only an extension.

“After their explorations have not Nordenskiold, Perry and Maaigaard stated that Greenland gets higher and higher towards the North Pole?

“Besides, they have found birds, different products and vegetables in the northern ice – ivory teeth also – which indicate that this region must have been inhabited and that animals must have been there, and perhaps people as well. There used to be large forests there, which must have been formed into coal-fields, which we will explore. Yes, there is a continent, without a doubt, around the North Pole – a continent free from all human beings, and on which we will place the banner of the United States.”

At this remark the auditors expressed great delight. When the noise had finally subsided Major Donellan could be heard to remark: "Well, seven minutes have already gone by of the ten which, as you say, would be sufficient to reach the North Pole."

"We shall be there in three minutes," coolly answered President Barbicane.

"But, even if this be a continent, which constitutes your purchase, and if it is a raised country, as we may have reasons to believe, it is also obstructed by eternal ice, and in a condition which will make exploration extremely difficult," responded the Major. "Impossible," cried Jan Harald, who emphasized this remark with a wave of his hand. "Impossible, all right," said Impey Barbicane. "But it is to conquer this impossibility that we have purchased this region. We will need neither vessels nor rafts to reach the North Pole; no, thanks to our operations, the ice and icebergs, new or old, will melt by themselves, and it will not cost one dollar of our capital nor one minute of our time." At this there was absolute silence. The most important moment had come.

“Gentlemen,” said the President of the Gun Club, “Archimedes only asked for a lever to lift the world. Well, this lever we have found. We are now in a position to remove the North Pole.”

“What, remove the North Pole?” cried Eric Baldenak.

“Will you bring it to America?” asked Jan Harald. Without doubt President Barbicane did not wish to explain himself just yet, for he continued: “In regard to this point of leverage –” “Do not tell it! do not tell it!” cried one of his associates, with a terrible voice.

“In regard to this lever–”

“Keep the secret! keep the secret!” cried the majority of the spectators, taking up the cry.

“We will keep it,” said President Barbicane.

Naturally, the European delegates were very much vexed at this remark. This will be easily understood. In spite of all these exclamations the orator never had any intention of making his plan known. He continued to say: “We obtained our object, thanks to a mechanical device, one which has no precedent in the annals of industrial art. We will undertake it and bring it to a successful finish by means of our capital, and how I will inform you forthwith."

"Hear! hear!" said the others present.

"First of all, the idea of our plan comes from one of the ablest, most devoted and illustrious calculators and one of our associates as well," said President Barbicane. "One to whom we owe all the calculations which allows us to have our work in such good condition. As the exploration of the North Pole is not a piece of play the removal of the pole is a problem which could only be solved by the highest calculations. Therefore we have called the assistance of the honorable Secretary, Mr. J.T. Maston."

"Hip, hip, hip, hurrah, for J.T. Maston," exclaimed all the auditors, seemingly electrified by the presence of this extraordinary calculator.

Mrs. Evangelina Scorbitt was deeply touched by this recognition of the celebrated mathematician, who had already entirely gained her heart. He contented himself with turning his head to the right and left, bowing and thanking his auditors.

“Already, dear subscribers,” said President Barbicane, “since the great meeting in honor of the arrival of the Frenchman, Michel Ardan, in America, some months before our departure for the moon” (and this confident Yankee spoke of the trip to the moon as quietly as if it were no more than a trip to New York), “J T. Maston had already said to himself: 'We must invent machines to move the North Pole. We must find a point for action and put the axis of the earth in the right direction from the object.' Well, any or all of you who listen to me find it if you can. I can only say the machines have been invented, the point of leverage has been found, and now let us pay our attention to the question of fixing, in the right way, for our end of the axis of the earth.” Here he stopped speaking, and the astonishment which was expressed on the faces of his auditors it is impossible to describe.

“What!” cried Major Donellan, “you then have the idea of putting the axis of the earth in another direction?”

“Yes, sir,” answered President Barbicane promptly. “We have the means of making a new one which will hereafter regulate the routine of day and night.”

“You want to modify the daily rotation of the earth?” repeated Col. Karkof, with fire in his eyes.

“Absolutely, but without affecting its duration,” answered President Barbicane. This operation will bring the pole at or about the sixty-seventh parallel of latitude, then the earth will be similar to the planet Jupiter, whose axis is nearly perpendicular to the plane of its orbit. Now this movement of 23 degrees 28 minutes will be sufficient to give at our North Pole such a degree of heat that it will melt in less than no time the icebergs and field which have been there for thousands of years.”

The audience was out of breath. Nobody thought of interrupting the orator, even to applaud him. All were taken in by this idea, so ingenious and simple, of modifying the axis on which this earthly spheroid is rotating. And as for the European delegates, well, they were simply stupefied, paralyzed, and crushed, they kept their mouths shut in the last stage of astonishment. But the hurrahs seemed to rend the hall asunder, when President Barbicane made the additional remark: “It is the sun which will take upon himself the melting of the icebergs and fields around the North Pole, and thus make access to the same very easy. So, as people cannot go to the pole, the pole will come to them.”

CHAPTER VIII

YES, JUST LIKE JUPITER.

***

Since that memorable meeting in honor of Michel Ardan, the Hon. J.T. Maston had talked and thought of nothing else but the "changing of the axis of the earth." He had studied the subject as much as possible and found out all the facts and figures about it. As the problem had been solved by this eminent calculator a new axis was going to take the place of the old one upon which the earth was now turning, and the world would otherwise remain the same. In the scheme it would be possible for the climate around the North Pole to become exactly the same as that of Trondhjem, in Norway, in the Spring. Naturally, then the large amount of ice would melt under the ardent sun. At the same time the climates would be distributed over our sphere like those now on the surface of Jupiter. In other words, the new-formed society of Barbicane was going to change everything at present on the surface of the earth. And the creation of this new axis was possible, just as soon as the platform of which Archimedes had dreamed and the lever imagined by J.T. Maston were at the disposal of these courageous engineers. And as they had decided to make a secret of their invention until a future time, people could not do anything else but make their own figures. This was what all the papers did, calling on the most scientific persons and learning as well something from the most ignorant persons. If there really were people living on the surface of Jupiter, they had a good many advantages over those on the earth, advantages which had all been narrated and explained in the meeting which was held before the trip to the moon. All these advantages would come to the people living on the earth if Barbicane & Co. could accomplish what they intended to do. Twenty-four hours would then always separate two noons from each other. Twilight and dawn would always be as they are now. But the most curious thing of all would be the absence of the different seasons of the year. Now there were Summer, Winter, Fall, and Spring. The people living on Jupiter did not know these seasons at all. After this experiment people living on this globe would not know them either. As soon as the new axis would be in smooth working order there would be no more ice regions, nor torrid zones, but the whole world would have an even temperature climate.

What, after all, is the torrid zone? It is a part of the surface in which the people can see the sun twice yearly at its zenith, and the temperate zone but a part where the sun never goes to the zenith, and the icy region but a part of the world which the sun forgets entirely for a long time, and around the North Pole this extends for six months. It is simply the position of the sun which makes a country exceedingly hot or cold. Well, these things would not appear any longer on the surface of the world. The sun would be always over the equator: it would go down every twelve hours just as regularly as before. "And among the advantages of the new method," said the friends of President Barbicane, "were these, that each person could choose a climate which was best for himself and his health; no more rheumatism, no more colds, no more grippe; the variations of extreme heat would not be known any more. In short, Barbicane & Co. were going to change fixtures which had existed ever since the world was in existence. Certainly the observer would lose a few stars and things which he perhaps liked to look at now, and the poet would not have any longer his dreamy nights, etc., but what a great advantage it would be for the world at large. “And,” said certain journals, “the products of the ground can be regulated so that agriculturists can give to each sort of plant life the temperature which suits it most.” Other newspapers asked: “Will we no more have rain, or storms, or hail – things upon which a great deal depends in the harvest time?” “Undoubtedly,” said the friends of Barbicane & Co., but these accidents will be more rare than they have been, as the temperature will be more even. Yes, taken in all, it will be a great advantage to humanity. It will be the real millennium of the earthly globe. And Barbicane & Co. will have done a service to mankind which but for them would have remained an impossibility.” “Yes,” said Michel Ardan, “our hemisphere, the surface of which is always either too cold or too warm, will no longer be the place for colds and rheumatism, etc.” A New York paper of Dec. 27 printed the following article: “Honor to President Barbicane! His associates and himself will not only annex a new province to our American continent, and thereby enlarge the already vast possessions of the United States, but they will make the whole world more productive and inhabitable. It will be possible then to put seed in the ground as soon as the crop had grown up and been taken out; there would be no more time lost during the Winter. And the coal mines also would make the country richer than the value of its entire present realty. Barbicane & Co. will change the whole world and put it in better condition. Thanks, then, to the people who have done this greatest of benefits to humanity."

CHAPTER IX

IN WHICH APPEARS THE FRENCH GENTLEMAN TO WHOM *WE REFERRED AT THE BEGINNING OF THIS* TRUTHFUL *STORY.*

***

Such, then, were to be the profits due to the changes which were to be wrought by President Barbicane. The earth would continue to revolve and the course of the year would not be much altered. As the changes would concern the whole world it was natural that they became of interest to all. In regard to the new axis which was going to be used that was the secret which neither President Barbicane nor Capt. Nicholl nor J.T. Maston seemed to be willing to give to the public. Were they to reveal it before, or would none know of it until after the change had taken place? A degree of uncertainty began to fill the American mind. Criticisms very natural and to be expected were made in the papers. By what mechanical means was this project to be carried out which would bring about this change? It would necessarily demand a terrible power. One of the greatest papers at that time commented in the following article: “If the earth was not turning on its axis, perhaps a very feeble shock would be sufficient to give it such a movement as might be chosen, but otherwise it would be very difficult if not impossible to deviate it a fixed amount.” Nothing seemed more correct after having discussed the effort which the engineers of the N.P.P.A. were to make. Discussion took on the interesting turn as to whether this result would be reached insensibly or suddenly. And if the latter, would not terrible accidents happen at the moment when the change took place? This troubled scientific people as well as ignorant people. It was not agreeable to know that a blow was to be struck and not know precisely what the after effects were to be.

It seemed as if the promoters of this undertaking had not fully considered the consequences – that they would be so very dangerous to the earth, and that it would not do as much good as first thought. The European delegates, more than ever angry at the loss which they had suffered, resolved to make the most of this question and to excite the public as much as possible upon it so as to turn feeling against the members of the Gun Club.

It will not be forgotten that France had absolutely nothing to do with these delegates, as it had no intentions of buying the Arctic region. However, a Frenchman had come to Baltimore, and for his own personal benefit and information had watched with great interest the proceedings of the Gun Club. He was an engineer, not more than thirty-five years old. He had been first in the polytechnic school, and came out of it with the highest honors. He was without doubt as skilful a calculator as Mr. J.T. Maston. This engineer was a very intelligent young man, very original, always pleasant, and with most amiable manners. He always spoke very frankly and used plain language, no matter whether he was speaking in earnest or in fun. He even went so far as to use slangy expressions when they served his purpose. He could sit for hours at his table and figure and calculate, making his figures and calculations as fast as he could write with a pen. His greatest pleasure, next to these difficult mathematical efforts, was in "whist," which he played apparently very indifferently, not forgetting to figure out all his chances. His name was Alcide Pierdeux, but he generally signed it, A. Pierd, and sometimes only A. Pie. He was very tall. His friends remarked that his height measured about the five millionth part of the quarter of the meridian, and they were not much mistaken. He had a small head, at least it looked so on his broad shoulders, but with a most lively expression on his face, and his blue eyes behind his eye-glasses twinkled merrily. This was characteristic of him, for he had one of those faces which appear merry, even when they are in sober earnest. He was at once the best scholar in his class and the best tempered. But even if his head did seem a little small on his shoulders, it is safe to say that it was filled to its highest capacity. He was a mathematician, as all his ancestors had been, but he did not study mathematics to use them in his profession, for which he never had any taste, as he disliked trade. No, he studied mathematics for themselves alone, simply to find them out more and more where there was so much unknown to man. Let us also remark that Alcide Pierdeux was a bachelor. He was as yet single, or, as he would express it, equal to one (= 1) although his greatest wish was to get married. His friends all thought that he would marry a very charming girl, gay and spirituelle. But, unhappily for him, the girl's father said that he was too smart and that he would talk to his daughter in language which she would not be able to understand. How modest and simple this father was, indeed. And for this reason the young engineer decided to place between himself and his country the broad ocean. He asked permission to go abroad for a year and obtained it. He thought that he could not make any better use of his time than to go to Baltimore and note the actions of the N.P.P.A. And this is how he came to be at this time in the United States. However, since he got to Baltimore he had cared little apparently for the great undertaking of Barbicane & Co. Whether the earth would have a change of the axis or not, what did it matter to him? He only wanted to know, and his curiosity was at the highest point to find out, by what means they were to move the earth. He thought again and again how they would do it and had several plans in his head and dismissed them only to consider the matter afresh. He concluded that they wanted probably to substitute a new axis, but he did not clearly see where their point of operations was to be. Then, again, he would say, "There is the daily movement. It is impossible to surpress it; how they will do it, is a perfect conundrum to me." He had no idea what the plans of Barbicane and Maston were. It is to be regretted very much that their intentions were not known to him, as he would have been able to figure out the formulae in a very short time. And so it came about that on this 29th day of December, Alcide Pierdeux was walking with his hand at his brow, pondering, about the streets of Baltimore.

CHAPTER X

CHAPTER X

***IN WHICH A LITTLE UNEASINESS BEGINS TO SHOW* ITSELF.**

***

A month had elapsed since the meeting of the Gun Club and the stockholders of the new-formed society, and public opinion was getting much altered. The advantages of the change to be wrought in the axis of the earth were forgotten and its disadvantages began to be spoken of. It was very probable, public opinion said, that a terrible catastrophe would happen, as the change could only be brought about by a violent shock. What would this catastrophe exactly be? In regard to the change of climates, was it so desirable after all? The Esquimaux and the Laps and the Samoyeden and the Tchuktchees would benefit by it, as they had nothing to lose. The European delegates were very energetic in their talk against President Barbicane and his work. To begin with they sent information to their Govern-

ment. They used the cable frequently and always sent cipher messages. They asked questions and received instructions. What, then, were these instructions, always in cipher and very guarded? "Show energy, but do not compromise our Government," said one. "Act very considerately, but do not touch the 'statu[s] quo,'" said another. Major Donellan and his associates did not fail to predict a terrible accident. "It is very evident that the American engineers have taken steps so as not to hurt, or at least as little as possible, the territory of the United States," thought Col. Boris Karkof. "But how could they do it?" asked Jan Harald. "If you shake a tree do not all its branches suffer while you are shaking it?" "And if somebody hits you on the back does not your whole body feel the pain?" said Jacques Jansen. "That is, then, what this strange paragraph of the document meant," said Dean Toodrink. "That is the reason why they mentioned certain geographical changes."

"Yes," said Eric Baldenak, "that is what we have to fear; this change will throw the sea out of its basin, and should the ocean leave its present quarters, would not certain inhabitants of this globe find themselves so located that they could not readily communicate with their fellow-citizens?"

"It is very possible that they may be brought into such a density of surrounding medium," said Jan Harald, gravely, "that they will be unable to breathe."

"We will see London at the top of Mount Blanc," exclaimed Major Donellan. And with his legs crossed and his head thrown back this gentleman looked straight up as if the capital of his country was already lost in the clouds. In short, it became a public danger and a most annoying one. True, it was only a question of a change of 23 degrees and 28 minutes, but this change might bring about a great movement of the oceans as the new earth flattened itself around the pole. Protestations were heard from all over, and the Government of the United States was asked to interfere. "It was best not to try the operation at all." "The consequences of it might destroy this world." "God has done all things well; it was not necessary to better his work," were the comments. And yet there were people light-hearted enough to make merry at the whole matter. "Look at these Yankees," they said, "they want to turn the earth on its axis. If the earth had shown any faults in its motion it would be all right to better it, but it had gone on for millions of years and always as regularly as clockwork."

Instead of answering such questions Engineer Alcide Pierdeux tried to find which would be the countries and directions, figured out by Mathematician Maston, in which the test would take place – the exact point of the globe where the work would begin. As soon as he should know this he would be master of the situation and know exactly the place which would be in the most danger. It has been mentioned before that the countries of the old continent were probably connected with those of the new across the North Pole. Was it not possible, it was asked in Europe, that President Barbicane and Capt. Nicholl and J.T. Maston had considered only how to save their own country from any ill consequences which might come from the shock? He was a Yankee – it was pointed out they were all Yankees – and particularly this man Barbicane, who had created the idea of going to the moon. In any case, it was argued, the whole new world, from the Arctic regions to the Gulf of Mexico, would not have to fear anything from the shock. It is even probable on the other hand that

America would profit immensely by it and gain some territory. “Who knows what is lying in the two oceans which wash the American coast? Was it not probable that there was some valuable territory which they wished to take possession of?” asked people who never saw anything but the dark side of a question. “Is it sure that there is no danger? Suppose J.T. Maston should make a mistake in his calculations? And could not the President have made a mistake when he came to put his apparatus in working order? This might happen to the smartest people. They might not always put the bullet in the target, or they might neglect to put the cannonball into the cannon,” were the comments of these nervous folk. This uneasiness was fomented by the European delegates. Secretary Dean Toodrink published several articles in this line, and even stronger ones were put by him in the *Standard*. Jan Harald put some in the Swedish journal *Aftenbladt*, and Col. Boris Karkof in a Russian journal which had a large circulation. Even in America opinions differed. The Republicans were friends of President Barbicane, but the Democrats declared themselves against him. A part of the American press agreed with the European press. And as in the United States the papers had become great powers, paying yearly for news about twenty millions of dollars, they had great influence on the people. In vain did other journals of large circulation speak in favor of the N.P.P.A. In vain did Mrs. Evangelina Scorbitt pay as high as $10 a line for articles showing the advantages of this invention. In vain did this ardent widow try to demonstrate that everything was perfectly correct, and that J.T. Maston could never commit an error in figuring. Finally America took fright in the matter and was inclined to be governed by Europe. But neither President Barbicane nor Secretary Maston of the Gun Club seemed to care what was said. They did not even take the trouble to correct the different articles. They let people say what they liked and did not try to change their minds at all. They were too much occupied in preparations for the immense undertaking. It is indeed strange that the public, who were at first so enthusiastic and so certain of success, should so suddenly turn and go against this operation.

Soon, however, in spite of the money Mrs. Evangelina Scorbitt spent on the matter, the President and Secretary of the Club came to be considered dangerous characters by the people of the two worlds. The Government of the United States was asked officially by the European powers to interfere and examine the matter. The originators were to openly show their ideas and by what means they hoped to accomplish what they intended. They would have to inform the Government which parts of the world would be most in danger and, in short, tell everything which the public demanded to know. The Government at Washington was compelled to do what they were asked. The uprising of public sentiment in the Northern, Southern, and Middle States of the Union did not allow them any other course. A commission of engineers, mechanicians, mathematicians, and geographers were appointed – fifty in all, presided over by John Prestice – by the act of the 19th of February, with full power to do anything which they considered necessary in the matter. At first the President of the Society received orders to appear before this committee. President Barbicane did not respond. Agents went to his house in Baltimore, but the President was gone. Where was he? No one knew. When did he depart? Six weeks ago, on the 11th of January, he had left the city, and the State of Maryland as well, in company with Capt. Nicholl.

Where did they both go? Nobody could tell. Evidently the two members of the Gun Club went to that mysterious region where preparations were going on for the great operation. But where could this place be? It was most important to know where this place was in order to break up and destroy the plans of these engineers before they had got too far in their work.

The consternation produced by this departure of the President and his associate was enormous. It soon changed public opinion to hatred against the N.P.P.A. and its managers. But there was one man who ought to know where the President and his associate had gone. There was one man who could answer this gigantic question, which at present excited the whole world and this man was – J.T. Maston. He was ordered to appear before the Committee of Inquiry under the Presidency of John Prestice. He did not appear. Had he also left Baltimore? Had he also gone to join his associates to aid them in their work, the results of which the whole world now expected with such immense fright? No. J.T. Maston was living still in his Ballistic Cottage, at No. 179 Franklin Street, working all the time and already beginning new calculations, only interrupting his work when he wanted to spend a social evening with Mrs. Evangelina Scorbitt at her magnificent residence at New Park. An agent was sent to him by the President of the Inquiry Committee with orders to bring him to their meeting. The agent arrived at the cottage, knocked at the door and introduced himself. He was harshly received by "Fire-Fire," but much worse by the proprietor of the house. However, Mr. Maston thought it was no more than right that he should go to the meeting, and he went with the agent. As soon as he had arrived they began to question him.

The first question was, "Where is President Barbicane and Capt. Nicholl at present?" He answered with a steady voice, "I know where they are, but I am not at liberty to disclose this information." Second question: "Have he and his associates made the necessary preparations to put this operation in working order?" "This," said Maston, "is a part of the secret which I cannot reveal." "Would he be land enough to let this Committee examine his own work, so they would be able to judge if his Society would be in position to accomplish their intentions?" "No, most certainly I shall not allow it, never; I would rather destroy it. It is my right as a citizen of free America to refuse to communicate to any person the result of my work."

"But," said President Prestice in a very serious voice, "if it is your right to keep silent, it is the right of the whole United States to ask you to stop these rumors and give an explanation of the means which will be employed by your Company," Mr. Maston did not agree that it was his right nor that it was his duty to answer further questions. In spite of their begging, threatening, etc., they could obtain nothing from this man with the iron hook. Never, never, would he say one word of it, and it was hardly possible to believe that such a strong will was concealed under that cover of "gutta-percha." Mr. Maston went away as he had come; he was congratulated by Mrs. Evangelina Scorbitt, who was delighted by the courageous attitude taken by him. When the results of this last meeting of the Inquiry Committee became known public indignation really took a turn which threatened the security and safety of the calculator. The pressure of public opinion was so great that the Cabinet of the Government of the United States was compelled to give the Committee full permission to do what they thought most necessary and advisable in the matter. One evening, the 13th of March, J.T. Maston was in his study at the Ballistic Cottage, very much interested in different figures, when suddenly the telephone bell attracted his attention. "Hello! hello!" said he, annoyed by this sudden interruption, "who wants me?" "Mme. Scorbitt." "What does Mrs. Scorbitt want?" "She wants to put you on your guard, I am informed this moment" – and she had not time to finish the phrase when Mr. Maston heard a terrible noise at the door of his house. On the stairs which led to his study there was an extraordinary racket. He could hear loud voices, many angry voices. Then the noise of a whole army of men moving towards his door. It was his servant Fire-Fire, who was trying to keep the intruders from breaking, into the house and disturbing the "home" of the master. A moment afterwards the door was violently opened and a policeman appeared, followed by several others. This policeman had a warrant to make a visit to the house and to take possession of all papers and also of J.T. Maston himself. The angry Secretary of the Gun Club reached for his revolver, and would have certainly defended himself had he not been suddenly disarmed. He was held by officers, and all his papers were put in a bundle. Suddenly he made a bold effort, freed himself, grabbed his note-book, out of which he tore the last page and began to chew it very quickly. "Now you can take it," said he, "for it will be no good to you." An hour afterwards he was a prisoner in the jail of Baltimore. This was undoubtedly the best that could happen to him, as it was extremely dangerous for him to be at liberty due to the then excited state of the public mind.

CHAPTER XI

WHAT WAS FOUND IN THE NOTEBOOK OF J.T. MASTON *AND WHAT IT NO LONGER CONTAINED.*

***

The notebook, which was taken possession of by the police, had thirty pages covered with formulae and figures, including all the calculations of J.T. Maston. It was a work of the higher mathematics, which could only be appreciated by the highest mathematicians. The following formula,

$$\frac{1}{2}\left(v^2 - v_0^2\right) = gr\left\{\frac{r}{x} - 1 + \frac{m'}{m}\left(\frac{r}{d-x} - \frac{r}{d-r}\right)\right\}$$

[Equation 1]

which was also to be found in the calculation of *From the Earth to the Moon,* held a prominent place in these calculations. The majority of people could not understand anything of what was written in the notebook, but it would have given satisfaction to give out the results, which every one expected with so much curiosity. And so it was that all the newspapers, and the Inquiry Committee as well, tried to read the formulae of this celebrated calculator. In the work of Mr. Maston were found some problems correctly executed, others half solved, etc. The calculations had been made with great exactness and of course the Inquiry Committee supposed that they were absolutely correct. If the plan was carried out fully it was seen that without a doubt the earth's axis would be greatly changed and that the terrible disasters which were predicted would take place with full force. The reports made by the Inquiry Committee to the different newspapers ran as follows:

> “The idea followed by the Administrative Council of the N.P.P.A. and the object of which is to substitute a new axis for the old one is to be carried out by means of the recoil of a piece of ordnance fixed at a certain point of the earth. If the barrel of this device is immovably fixed to the ground it is not at all doubtful that it will communicate its shock over our whole planet. The engine adopted by the engineers of the Society is then nothing else but a monster cannon, the effect of shooting which would be absolutely nothing if it were pointed

> vertically. To produce its highest effect it is necessary to point it horizontally towards the north or south, and it is this last direction which has been chosen by Barbicane & Co. Under these conditions the recoil will produce a movement of the earth towards the north, a movement similar to that of one billiard ball touched very slightly by another."

This was really just what the clever Alcide Pierdeux had predicted. As soon as the cannon has been fired off, the center line of the earth would be displaced in a parallel direction to that of the recoil. This would change the direction of the orbit somewhat, and consequently the duration of the year, but in such a mild way that it must be considered as absolutely free from bad results. At the same time the earth takes a new movement of rotation around an axis in the plane of the equator, and the daily rotation will then be accomplished indefinitely upon this new axis, as if no daily movement had existed previous to the shock. At present this movement is made around the lines of the poles, and in combination with the accessory force produced by the recoil there was created a new axis, the pole of which moves from the present to the amount of a quantity called "x." In other words, if the cannon is fired at the moment when the vernal equinox – one of the intersections of the equator and the ecliptic – is at the nadir of the point of shooting, and if the recoil is sufficiently strong to displace the old pole 23 degrees, 28 minutes, the new axis becomes perpendicular to the direction of the earth's orbit, the same as it is for the planet Jupiter.

What the consequences were expected to be we already know, as President Barbicane had indicated them at the meeting of the 22d of December. But, given the mass of the earth and the quantity of momentum, which she possesses, is it possible to conceive a piece of ordnance so strong that its recoil will be able to produce a modification in the actual direction of the real pole, and especially to the extent of 23 degrees, 28 minutes? Yes, if a cannon or a series of cannons are built with the dimensions required by the laws of mechanics, or, in lieu of these dimensions, if the inventors were in possession of an explosive strong enough to impel a projectile with the necessary velocity for such a displacement.

Now, taking as a basis model the cannon of 27 centimetres of the French Marine Corps, which throws a projectile of 180 kilograms with an initial velocity of 500 metres a second, by giving to this piece of ordnance an increased dimension of 100 times – that is, a million times in volume – it would throw a projectile of 180,000 tons: or, in other words, if the powder had strength sufficient to give to the projectile an initial velocity 5,600 times greater than that of the old black powder used for a cannon the desired result would be obtained. In fact, with a velocity of 2,800 kilometres a second, a velocity sufficient to go from Paris to St. Petersburg in one second, there was no doubt that the recoil of the projectile, acting against the earth, would put everything again in a state of quietude. Well, extraordinary as it may appear, J.T. Maston and his associates had in their possession exactly this explosive, of a nearly unlimited power, and of which the gunpowder used to throw the ball of the [C]olumbiad towards the moon gave but a faint idea. It was Capt. Nicholl who had discovered it. The substances which entered into its composition were only imperfectly entered in the notebook of Mr. Maston, and he merely named it "melimelonite." All that was known was that it was formed by the reaction of a melimelo of organic substances and azotic acids. No matter what the explosive was, with the power which it possessed it was more than sufficient to throw a projectile weighing 180,000 tons outside of the earth's attraction, and it was evident that the recoil which it would produce to the cannon would have the effect of changing the axis, displacing the North Pole 23 degrees and 28 minutes, bringing the new axis in the direction of the ecliptic, and, as a consequence of this, effecting all the changes so justly dreaded by the inhabitants of the earth.

However, there was one chance for humanity to escape the consequences of this trial, which was to provoke such revulsions in the geographical and climatic conditions of the globe. Was it possible to build a cannon of such dimensions that it was to be a million times greater in volume than the one of 27 centimetres? It was doubtful. That was just the point and one of the reasons for thinking the attempt of Barbicane & Co. would not succeed. But there was the other possibility, for it seemed that the Company had already begun to work on their gigantic project. Now the question arose, where was their place of operations? No one knew, and consequently it was impossible to overtake these audacious operations. It was well known that Barbicane and Nicholl had left Baltimore and

America. They had gone away two months ago. Where were they? Most certainly at that unknown point of the globe where the operations were under way for their grand object. It was evident that this place was indicated on the last page of the note-book of J.T. Maston. On this point there was no doubt. But this last page had been torn out and eaten up by the accomplice of Impey Barbicane, and Maston sat imprisoned in the Baltimore City Prison and absolutely refused to speak. This was the condition of affairs. If the President succeeded in making this monster cannon and its projectile – in a word, if the operation was carried out under the above stated conditions – it would modify the earth's axis, and within six months the earth would be subject to the consequences of this audacious attempt of Barbicane & Co. This would come on the 22d day of September, twelve hours after the passage of the sun over the meridian of the place "x."

The facts that were known were: 1st. That the shooting would be done with a cannon a million times larger than the cannon of 27 centimetres. 2d. That the cannon would be loaded with a projectile of 180,000 tons. 3d. That the projectile would be animated with a velocity of 2,800 kilometres.

4th. That the shooting would take place on the 22d of September, twelve hours after the passage of the sun over the meridian of the place “x.” Was it possible to deduce, under these facts, where was the spot “x,” where the operation was to take place? Evidently not, said the Inquiry Committee. There was nothing by which to calculate where the point “x” was, as nothing in the calculations of Mr. Maston indicated through which point of the globe the new axis was to pass, or, in other words, on which part of the present earth the new poles would be situated. Therefore, it would be impossible to know which would be the elevated and submerged countries, due to the changed surface of the ocean, or which parts of the earth would be transformed into water, and where water would be transformed into land. It was evident that the maximum change in the ocean surface would be 8.415 metres, and that in certain points of the globe various areas would be lowered and raised to this amount. All, however, depended upon the location of the point “x,” or where the shooting was to take place. In other words, “x” was the secret of the promoter of this uncertain affair. “We have,” said the Committee, “only to mention again that the inhabitants of the world, no matter in what part of it they are living, are directly interested in knowing this secret, as they are all directly t[h]reatened by the actions of Barbicane & Co. Therefore all the inhabitants of Europe, Africa, Asia, America, and Australia are advised to watch all gun foundries, powder factories, etc., which are situated in their territory and to note the presence of all strangers whose arrival may appear suspicious, and to advise the Inquiry Committee at Baltimore by wire immediately. Heaven grant that this news may arrive before the 22d of September of the present year, as that date threatens to disturb the order established since the creation in our earthly system.

CHAPTER XII

CHAPTER XII

IN WHICH J.T. MASTON HEROICALLY CONTINUES TO BE SILENT.

***

According to a former story a gun was to be employed to throw the projectile from the earth to the moon; now the gun was to be employed to change the earth's axis. The cannon, always the cannon; these gunners of the Gun Club had nothing else in their heads but the cannon. They had a real craze for the cannon. Was this brutal engine again threatening the universe? Yes, we are sorry to confess it, it was a cannon which was uppermost in the mind of President Barbicane and his associates. After the Columbiad of Florida, they had gone on to the monster cannon of the place "x." We may almost hear them shout with a loud voice: "Take aim at the moon." First act, "Fire." "Change the axis of the earth." Second act, "Fire." And the wish which the whole world had for them was, "To hell." Third act, "Fire." And really their scheme justified the popular opinion.

As it was, the publication of this last report of the Committee in the newspapers produced an effect of which one can scarcely form an ideal. The operation to be tried by President Barbicane and Capt. Nicholl, it was very clear, was going to bring about one of the most disastrous interruptions in the daily routine of the earth. Everybody understood what the consequences of it would be. Therefore the experiment of Barbicane & Co. was generally cursed, denounced, etc. In the Old as well as in the New World the members of the N.P.P.A. had at the time only enemies. If there were indeed a few friends left to them among their cranky American admirers, they were very few.

Regarding only their personal security, President Barbicane and Capt. Nicholl had acted wisely in leaving Baltimore and America. It was safe to believe that some accident had happened to them. They could not without divine punishment threaten fourteen hundred million inhabitants by a change wrought in the habitability of the earth.

But how was it possible that the two leaders of the Gun Club had disappeared without leaving any trace behind them? How could they have sent away the material and assistants which were necessary to such an operation without any one seeing them? A hundred railroad cars, if it was by rail, a hundred vessels, if it was by water, would not have been more than sufficient to transport the loads of metal of coal, and of me-limelonite. It was entirely incomprehensible how this departure could have been made incognito. However, it was done. And still more serious it appeared when it was known after inquiry that no orders had been sent to the gun foundries or powder factories, or the factories which produce chemical products in either of the two continents. How inexplicable all this was! Without doubt it would be explained some day.

At any rate, if President Barbicane and Capt. Nicholl, who had mysteriously disappeared, were sheltered from any immediate danger, their colleague, Mr. Maston, was under lock and key, and had to face all the public indignation. Nothing could make him yield, however. Deep at the bottom of the cell which he occupied in the prison of Baltimore, the Secretary of the Gun Club gave himself up more and more to thinking of those distant associates whom he was not able to follow. He pictured the vision of President Barbicane and his associate, Capt. Nicholl, preparing their gigantic operation at this unknown point of the globe, with nothing in their way. He saw them build their enormous device, combining their melimelonite, moulding the projectile which the sun would so soon count as one of its small satellites. This new star was to have the charming name "Scorbetta," in gallant acknowledgment of the love and esteem felt towards the rich capitalist widow of New Park. J.T. Maston calculated the days which would elapse before the one on which the gun would be fired.

It was already the beginning of April. In two months and a half the meridian star, after having stopped on the Tropic of Cancer, would go back towards the Tropic of Capricorn. Three months later it would traverse the equatorial line at the Fall equinox.

And then these seasons, which have appeared annually for millions of years, and which have changed so regularly, will be brought to an end. For the last time in 189– the sphere would have submitted to this succession of days and nights.

Truly, this was a magnificent work, superhuman, even divine. J.T. Maston forgot the Arctic region and the exploration of the coal mines around the pole, and he only saw, in his mind's eye, the cosmographic consequences of the operation. The principal object of the association was now to make those changes and displacements which were to remodel the face of the earth.

But that was just the point. Did the earth wish to change her face at all? Was she not still young and charming with the one which God had given her at the first hour of her creation?

Alone and defenseless in his prison cell, nothing could induce Mr. Maston to speak about the matter, no matter what plan was tried. The members of the Inquiry Committee urged him daily to speak, and visited him daily, but they could obtain nothing. It was about this time that John Prestice had the idea of using an influence which might possibly succeed, and this was the aid of Mrs. Evangelina Scorbitt. Every one knew what feelings the generous widow entertained for Mr. Maston, how devoted she was to him, and what unlimited interest she had in this celebrated calculator. Therefore, after deliberation of the Committee,

Mrs. Evangelina Scorbitt was authorized to come and go, visiting the prisoner as much as she liked.

Was she not threatened just as well as any other person on this earth by the recoil of this monster cannon? Would her palace at New Park be spared any more than the smallest hut of the Indian? Was not her very existence just as much in doubt as that of the savage living on the furthest isle of the Pacific Ocean? That is what the President of the Inquiry Committee gave her to understand, and for this reason she was begged to use her influence with the mathematician. If he would consent to speak, and would say at what place President Barbicane and Capt. Nicholl were, and how many people they had with them to accomplish their ends, it would yet be time to go and stop them and put an end to their project, and thus save humanity from this most dangerous catastrophe which threatened the world.

Mrs. Evangelina Scorbitt was therefore admitted to the prison whenever she wished it. She was most desirous of seeing J.T. Maston again after he had been taken from his comfortable study at Ballistic Cottage by those rough police agents. If any impolite person had on the 9th of April put his ear at the door of his cell the first time when Mrs. Scorbitt entered he would have heard the following conversation:

"Ah, at last, my dear Maston, I see you again."

"You, Mrs. Scorbitt!"

"Yes, my dear friend, after four weeks – four long weeks of separation."

"Exactly twenty-eight days, five hours and forty-five minutes," answered J.T. Maston, after having consulted his watch.

"Finally we are reunited."

"But how did it happen that they allowed you to penetrate as far as this cell to see me, dear Mrs. Scorbitt?"

"Under the condition of using all my influence over you, thanks to my affection for you, in advising you to disclose the secret of the whereabouts of President Barbicane."

“What, Evangelina!” cried Mr. Maston, “and you have consented to give me such advice. You have entertained the thought that I could betray my associates.”

“Me, dear Maston! Do you consider me so bad? Me! To sacrifice your security for your honor. Me! To persuade you to an act which would shame a life consecrated entirely to the highest speculations of pure mathematics.”

“Bravo, Mrs. Scorbitt! I see in you once more the generous patron of our Society. No, I have never doubted your great heart.”

“Thank you, Mr. Maston.”

“In regard to myself,” continued Maston, “allow me to say, before telling the point of the earth where our great shooting will take place – sell, so to speak, the secret which I have been able to keep so well, to allow these barbarians to fly and pursue our friends, to interrupt their works, which will make our profit and glory, I would rather die.”

“Splendid, Mr. Maston!” cried Mrs. Evangelina Scorbitt.

And these two beings, united by the same enthusiasm, crazed by it if you will, one as well as the other, were well matched in understanding each other perfectly.

“No, they will never know the name of the country which my calculations have designated, and the reputation of which will become immortal,” said J.T. Maston. “They can silence me if they like, but they will never have the secret from me.”

“And they can kill me with you,” said Mrs. Evangelina Scorbitt; “I will also be mute.”

“It is lucky, dear Evangelina, that they are ignorant of your knowledge of the place.”

“Do you believe that I would be capable of betraying it, because I am only a woman? Betray my associates and you! No, my friend, no. If they should raise the whole city and country against you – if the whole world would come to the door of this cell to take you away, I shall be there, too, and we will at least have one consolation – we will die together.”

As if there could be any greater consolation and Mr. Maston could dream of a sweeter death than dying in the arms of Mrs. Evangelina Scorbitt! And so ended the conversation every time that this excellent woman visited the prisoner. And when the Inquiry Committee asked her what the result was, she would say: “Nothing as yet; perhaps with time I shall be able to reach my point.”

Ah, women, women! What are women? “In time,” she urged. But time went on with fast steps. Weeks went 'round like days, days like hours and hours like minutes.

It was already May. Mrs. Evangelina Scorbitt had not been able to get any information from J.T. Maston, and where she had failed there was no hope of any other person succeeding.

Was it, then, necessary to accept this terrible shock without interfering in any way? No, no! Under such circumstances resignation was impossible. The European delegates became more and more out of spirits. There was wrangling between them every day. Even Jacques Jansen woke up out of his Dutch placidity and annoyed his colleagues greatly by his daily charges and countercharges.

Col. Boris Karkof even had a duel with the Secretary of the Inquiry Committee in which he only slightly injured his adversary. And Major Donellan; well, he neither fought with firearms nor with bare fists, quite contrary to English use, and he only looked on while his Secretary, Dean Toodrink, exchanged a few blows according to prize-ring rules with William S. Forster, the phlegmatic dealer in codfish, the straw man of the N.P.P.A., who really knew absolutely nothing of the affair.

The whole world was leagued against the United States and wanted to hold the Americans responsible for the actions of one of their number – the celebrated Impey Barbicane. There was talk of recalling the ambassadors and the foreign Ministers at present accredited to this most reckless Government at Washington and of declaring war against the United States. Poor United States! It only wished to lay its hands on Barbicane & Co. In vain did the Republic reply to the Powers of Europe, Asia, Africa and Australia that they were at liberty to arrest these adventurous Americans wherever they found them. Nobody would listen patiently to such talk. And so, far away President Barbicane and his associate were occupied in preparing their great operation. As nothing could be found of them the foreign countries began to say: “You have their accomplice; now it is sure that Mr. Maston knows where these people are and what they are doing. Make him speak, this man, Mr. Maston. Why not use hot oil, melted lead, etc.? Why not use such means as were used formerly under circumstances less grave and for cases which only interested a few private people? But it was answered that, while such means were justified in former times, they could not be used at the end of a century as far advanced as the nineteenth century was. Therefore, J.T. Maston had nothing to fear in that line; all that was left to hope was that he would finally consider the enormity of his crime and would decide to reveal his secret, or that some accident would reveal it for him.

CHAPTER XIII

AT THE CLOSE OF WHICH J.T. MASTON UTTERS AN EPIGRAM.

***

Time went on, however, and very likely also the works of Barbicane and Capt. Nicholl who were going on also under these very surprising conditions, no one knew where.

How was it possible, it was asked, that an operation which required the establishment of a considerable iron foundry, the erection of high blast furnaces, capable of melting a mass of metal a million times as large as the marine corps cannon of 27 centimeters, and a projectile weighing 180,000 tons, all of which necessitated the employment of several thousand workmen, their transport, their management, etc., – yes, how was it possible that such an operation could go on without the interested world getting any knowledge of it. In which part of the Old or New World had

Barbicane & Co. secretly established a foothold so that no hint was given to people living in the vicinity? Was it on a deserted island in the Pacific Ocean or in the Indian Ocean? But there were no more deserted islands: the English had gobbled them all up. Perhaps the new Society had discovered one for this special purpose. Perhaps, one remarked, they might be in some part of the arctic regions. No, this could not be, as it was simply because they could not be reached that the N.P.P.A. was going to remove them. Therefore, to look for President Barbicane and Capt. Nicholl on one of these islands or in some inaccessible point was simply wasting time. Did not the notebook taken away from J.T. Maston state that the shooting would take place on or about the equator? And all the countries around it were inhabited by some people. It seemed impossible for them to be so secreted in any part of the habitable world without some one informing the committee at Baltimore.

Now, what did Alcide Pierdeux think of all this? He was dreaming of all kinds of consequences which this operation would have. That Capt. Nicholl had invented an explosive of such tremendous power, that he had found the melimelonite, with an expansive force three or four thousand times stronger than that of the most violent explosive known, and 5,600 times stronger than the good old black gunpowder of our ancestors, this was astonishing enough – very astonishing. But it was not impossible at all. One can hardly know what the future will bring in these days of progress when devices exist to destroy whole armies at very long distances. In any event, the change of the earth's axis, produced by the recoil of a piece of ordnance, was not sufficiently novel to astonish the French engineer. Then, considering the plans of President Barbicane, he said: "It is evident that the earth receives daily the recoil of all the blows which are given on its surface. Hundreds of thousands of people amuse themselves daily by sending thousands of projectiles weighing a few kilograms or millions of projectiles weighing a few grammes, and even when I walk or jump, or when I stretch out my arm, all this takes place on the surface of our sphere and adds to or checks its motion. Is, then, your great machine of such a nature as to produce the recoil asked for? How in the name of candor can this recoil be sufficient to move the earth? And if the calculations of this fellow, J.T. Maston, prove it, it is easy enough to show it. Alcide Pierdeux could not but admire the ingenious calculations of the Secretary of the Gun Club, which were communicated by the members of the Inquiry Committee to those wise people who were able to understand them. And Alcide Pierdeux, who was able to read algebra like one would read a newspaper, found in this sort of reading matter an inexpressible charm. If these changes were to take place, what a terrible catastrophe it would be! Towns would be turned upside down, oceans would be thrown out of their beds, people killed by millions. It would be an earthquake of incomparable violence. If besides, said Alcide Pierdeux, this damnable powder of Capt. Nicholl were less strong, we might hope that the projectile would again strike the earth after the shooting, and after having made the trip around the globe, then everything would be replaced in a very short time and without having caused any very great destruction. But do not worry about that. Thanks to their melimelonite, the bullet will go its way and not return to the earth to beg her pardon for having deranged her by putting her back again in her place. Pierdeux finally said: "If the place of shooting were known I would soon be able to say upon which places the movement would have the least and where the greatest effect. The people might be informed in time to save themselves before their cities and houses had fallen under the blow." But how were we to know it? "I think," he said, "the consequences of the shock may be more complicated than can even be imagined. The volcanoes, profiting by this occasion, would vomit like a person who is seasick. Perhaps a part of the ocean might fall into one of their craters. It would make small difference then. It is entirely possible that we might have explosions which would make our earth jump. Ah, this Satan Maston, imagine him juggling with our earthly globe and playing with it as if he were playing billiards!"

So talked and reasoned Alcide Pierdeux. Soon these terrible hypotheses were taken up and discussed by the newspapers. The confusion which would be the result of the scheme of Barbicane & Co. could only result in terrible accidents. And so it happened that the nearer the day came the greater the fright which took possession of the bravest people. It was the same as it was in the year 1000, when all living people supposed that they would be thrown suddenly into the jaws of death. It maybe recalled what happened at this period. According to the Apocalypse the people were led to believe that the judgment day had come. In the last year of the 10th century, says H. Martin, everything was interrupted – pleasures, business, interest, all, even the public works of the country. Thinking only of the eternity which was to begin on the morrow, provision was made only for the most necessary articles for one or two days. All possessions, real estate, castles, were bequeathed to the Church, so as to acquire protection in that kingdom of heaven where all were so soon to enter. Many donations to the churches were made with these words:

"As the end of the world has come, and its ruin is imminent."

When this fatal time came, all the people ran to the churches and places set apart for religious meetings, and waited to hear the seven trumpets of the seven angels of the judgment day sound and call from heaven. We know that the first day of 1,000 came and went, and nothing was changed. But this time it was not the question of a disturbance simply based upon some verse of the Bible. It was the question of removing the axis of the earth, and this was founded on very reliable calculations, and was very probable.

Under these conditions the situation of J.T. Maston became each day more and more critical. Mrs. Evangelina Scorbitt trembled lest he would become the victim of a universal cry for vengeance. Perhaps she even had in her mind the idea of making him give up the information which he so heroically held to himself. But she did not dare to mention it to him and she did well. It would have been unwise for her to expose herself to the volley of rebukes he would have given her. As we may well understand, fright had taken a strong foothold in the city of Baltimore and the inhabitants became nearly unmanageable. The excitement was increased by articles appearing in the daily papers. In any case, if J.T. Maston had been found among the crowd of people, his fate would have been soon settled. He would have been given to the wild beast. But he was content and said: “I am ready for it.” No matter what happened, J.T. Maston refused to make known the situation of the “x,” knowing very well that if he should unveil the secret President Barbicane and Capt. Nicholl would be unable to finish their work. It was an interesting struggle – this fight of one man against the whole world. It only made J.T. Maston a grander and better man in the eyes of

Evangelina Scorbitt, and also in the opinion of his associates of the Gun Club. The Secretary of the Gun Club became such a celebrated person that he began to receive letters, as all criminals do, from people who wished to have a few lines from the hand which was going to turn the world over. But even if this was all very nice it became every day more and more dangerous for our Secretary. The population hung day and night around the prison, with great noise and great tumult. The enraged crowd wanted to lynch J.T. Maston. The police saw the moment would come when they would be unable to defend the prison and the prisoner J.T. Maston. Being desirous of giving satisfaction and information to the American people, as well as to the people of other countries, the Government at Washington decided to put J.T. Maston before a court of justice. “What other people have not been able to accomplish the Judges will not,” said Alcide Pierdeux, who had after all a kind of a friendly feeling for the unhappy calculator.

On the morning of the 5th September the President of the Commission went personally to the cell of the prisoner. Mrs. Evangelina Scorbitt, at her own request, had been allowed to accompany him. Perhaps at this last attempt the influence of this excellent lady would succeed and bring the hoped-for result. There was nothing to be left undone. All means possible were to be used to make this last attempt successful. If it was not – well, we will see. "Yes, we will see." What we would see is the hanging of this brute Maston, said the people, and the event would have come off in all its horror if the people could have it their way. So it happened that at 11 o'clock J.T. Maston was ushered into the presence of Mrs. Evangelina Scorbitt and John Prestice, President of the Inquiry Committee.

The opening was a very simple one. The conversation consisted of the following questions and answers, very rapid on one side and very quiet on the other. And even under these circumstances the calm, quiet speaker was J.T. Maston.

"For the last time will you answer?" asked John Prestice.

"Answer what?" ironically observed the Secretary of the Gun Club.

“Answer the question, where is the place in which your associate, Barbicane, is at present.”

“I have told it to you a hundred times.”

“Repeat it for the one hundred and first time.”

“He is where the shooting will take place.”

“Where will the shooting take place?”

“Where my associate, Barbicane, is.”

“Have a care, J.T. Maston.”

“For what?”

“For the consequences of your refusal to answer, the result of which will be–”

“To prevent you from learning that which you should not know.”

“What we have the right to know.”

“That is not my opinion.”

“We will bring you before the court.”

“Go ahead.”

“And the jury will condemn you.”

“What care I.”

“And as soon as judgment is rendered it will be executed.”

“All right.”

“Dear Maston,” ventured Mrs. Evangelina Scorbitt, whose heart nearly broke on account of these terrible threats.

“What! You, madam?” said J.T. Maston.

She hung her head and was silent.

“And do you want to know what this judgment will be?”

“If you wish to tell it,” said J.T. Maston.

“That you will suffer capital punishment, as you deserve.”

“Really?”

“That you will be hanged as sure, sir, as two and two make four.”

“Then, sir, I have yet a chance,” said J.T. Maston, reflectingly. “If you were a little better mathematician you would not say that two and two are four. You simply prove that all mathematicians have been fools until to-day in affirming that the sum of two numbers is equal to one of their parts; that is, two and two are exactly four.”

“Sir!” cried the President, absolutely puzzled.

“Well,” said J.T. Maston, “if you would say, as sure as one and one are two, all right. That is absolutely evident, because that is no longer a theorem; this is a definition.”

After this lesson in simple arithmetic the President of the Committee went out, followed by Mrs. Evangelina Scorbitt, who had so much admiration

for the calculator that she did not venture to look at him.

CHAPTER XIV

VERY SHORT, BUT IN WHICH "X" TAKES A GEOGRAPHICAL *VALUE.*

***

Very luckily for J.T. Maston, the Federal Government received the following telegram sent by the American Consul stationed at Zanzibar:

> "To John S. Wright, Minister of State, Washington, U.S.A.:
>
> "Zanzibar, Sept. 13, 5 A.M. (local time). – Great works are being executed in the Wamasai, south of the chain of Kilimanjaro. For eight months President Barbicane and Capt. Nicholl have been established there with a great number of black help under the authority of Sultan Bali-Bali. This is brought to the knowledge of the Government by its devoted
>
> "RICHARD W. TRUST, Consul"

And this was how the secret of J.T. Maston became known. And therefore, were the Secretary of the Gun Club still in prison, he could not have been hanged.

But, after all, who knows whether he would not rather have been glad to meet with death in the full glory of his life than to live on with all the chances of disappointment.

CHAPTER XV

WHICH CONTAINS A FEW INTERESTING DETAILS FOR THE INHABITANTS OF THE EARTHLY SPHERE.

***

Finally the Government of Washington had found out the place where Barbicane & Co. were operating. Should they doubt the authenticity of this cable? No, that was not reasonable. The Consul at Zanzibar was a very reliable person, and his information could be accepted without doubt. It was further corroborated by later telegrams. It was really in the center of the region of Kilimanjaro in the African Wamasai, a little under the equatorial line, where the engineers of the N.P.P.A. were going to accomplish their gigantic works.

How could they have secretly reached this lost country, at the foot of the celebrated mountain, discovered in 1849 by Drs. Rebviani and Krapf, ascended by the travellers Otto Ehlers and Abbot? How were they able to establish their workshops there, erect a foundry and bring a large number of help, or at least enough to succeed? How had they been able to establish friendly relations with the dangerous tribes of the country and their sover[e]igns, as cunning as they were cruel? This we do not know. And perhaps it would never be known, as there were only a few days left before the 22d of September would arrive. J.T. Maston heard from Mrs. Evangelina Scorbitt that the mystery of Kilimanjaro had been unveiled by a telegram sent from Zanzibar. “Great Scott!” he exclaimed, sawing the air with his iron hand. “Well, we do not travel by telegram yet, nor by the telephone, and in six days the matter will be finished.”

Those who saw and heard this remarkable man utter these words were astonished at the energy in the old gunner.

J.T. Maston was right. There was no time left to send agents to Wamasai with orders to arrest President Barbicane. They would even have been too late had they departed from Algiers or Egypt, even from Aden, Madagascar, or Zanzibar, as they would have met thousands of difficulties in this mountainous region, and perhaps they would have met with an army composed of followers of the Sultan, who was interested in the matter. Therefore all hope of preventing this operation had to be given up. But if prevention was impossible nothing seemed more easy than the figuring out of the terrible consequences, as the exact situation of "x" was now known.

This problem was difficult enough, but all algebraists and mathematicians of large reputation ought to be able to solve it. As the cable of the Consul of Zanzibar had been sent direct to the Minister of State at Washington, the Federal Government wanted to keep it secret at first. They wished as well that its contents were published all over the country, so that they could indicate what the results would be of this displacement of the axis and the uprising of the oceans, and thus the inhabitants of the world might learn which place of refuge was open to them according to the section of the globe in which they lived. And it is easy to understand how anxious the people were to learn their fate.

On the 14th of September the cable dispatch was sent to the office of the Observatory at Washington, with orders to figure out the final consequences upon geographical locations. Two days afterwards the problem was all worked out. The Old World was notified of the results by cable and the New World by telegram. After this calculation had been published by thousands of papers, it was the only thing talked of in the great cities and everywhere. What will happen?

This was the question which everybody was asking at every point of the globe.

The following was the notice made by the Observatory at Washington:

IMPORTANT NOTICE

> The operation which is being tried by President Barbicane and Capt. Nicholl is as follows:
>
> The production of a recoil, on the 22d of September, at midnight, by means of a cannon a million times larger in volume than the cannon of twenty-seven centimetres, throwing a projectile of 180,000 tons, with a powder giving it a velocity of 2,800 kilometres.

Now, if this shooting takes place below the equatorial line, nearly on the thirty-fourth degree of latitude west of the meridian of Paris, at the foot of Kilimanjaro, and if it is directed towards the south, these are the mechanical effects which it will have on the earth's sphere: Instantly, in consequence of the shock acting with the daily movement a new axis will be formed and, as the old axis will be displaced to the amount of twenty-three degrees and twenty-eight minutes, according to the figures obtained by J.T. Maston, the new axis will be perpendicular to the direction of the ediptic.

Which point will the new axis start from? As the point of shooting is known, it has been easy to calculate this.

In the North the extremity of the new axis will be situated between Greenland and Grinnel-land, exactly on that part of Baffin's Sea where it cuts the Arctic polar circle. In the South it will be on the line of the antarctic circle, a few degrees east of Adelialand. Under these conditions a new zero meridian, starting from the new North Pole, will pass through Dublin in

Ireland, Paris in France, Palermo in Sicily, the Gulf of Grand Sytre on the coast of Tripoli, Obed in Darfur, the mountain chain of Kilimanjaro, Madagascar; the Kerguelen Island, in the Central Pacific; the new antarctic pole, the antipodes of Paris, Cook Island, the Island of Quadra, Vancouver, on the margin of British Columbia; across North America to Melville Island, in the neighborhood of the North Pole.

In connection with this new axis of rotation, starting from Baffins' Bay in the north, to Adelialand in the south, a new equator will be formed above which the sun will travel without ever changing his daily course. The equinoctial line will cross the Kilimanjaro, at Wamasai, the Indian Ocean, Goa and Chicacola, a little below Calcutta in India, Mandalay in the Kingdom of Siam, Kesho in Tonquin, Hong Kong in China, Risa Island, Marshall Island, Gaspar Rico, Walker Island in the Pacific, the Cordilleras in the Argentine Republic, Rio de Janeiro in Brazil, the islands of Trinity and St. Helena in the Atlantic Ocean, St. Paul de Loando on the Congo, and finally it will meet again in the territories of Wamasai, back of Kilimanjaro. This new equator being thus determined by the creation of the new axis, it became possible to calculate the changes of the ocean tides, which was so important for the security of the inhabitants of the earth. It is just to observe that the directors of the North Polar Practical Association had taken measure to weaken the shock as much as possible. If the shooting had been towards the north the consequences of it would have been much more disastrous for the more civilized parts of the earth. On the other hand, shooting towards the south the consequences would only be felt most in parts less populated and less civilized. The careful calculations made showed how the waters would be distributed when thrown out of their beds by the flattening of the sphere around the new poles. The globe would be divided by two great circles, intersecting in a right angle at Kilimanjaro, and at its antipodes in the equinoctial ocean. This would form four sections, two in the north and two in the south, separated by the lines upon which the ocean upheaval would be zero.

In the northern hemisphere: The first section west of Kilimanjaro would take in Africa from the Congo to Egypt, Europe from Turkey to

Greenland, America from English Columbia to Peru, and from Brazil as high as San Salvador, and finally the whole northern Atlantic Ocean and the largest part of the temperate Atlantic zone.

The second section, east of Kilimanjaro, would include the greater patt of Europe, from the Black Sea to Sweden, European and Asiatic Russia, Arabia, nearly the whole of India, Persia, Beloochistan, Afganistan, Turkestan, the Celestial Empire, Mongolia, Japan, Corea, the Black Sea, the Caspian Sea, the greater part of the Pacific Ocean, the territories of Alaska in North America, and also the polar region which belonged to the American society, North Polar Practical Association.

The southern hemisphere would embrace the third section east of Kilimanjaro, which would include Madagascar, the islands of Marion, Kerguelen, Maurice, Reunion, and all the islands of the Indian Ocean, the Antarctic Ocean (as far as the new pole), half the island of Malacca, Java, Sumatra, Borneo, the islands of Sonde, the Philippines, Australia, New Zealand, New Guinea, New Caledonia, all the northern parts of the Pacific and its numerous archipelagos, nearly up to the 160th meridian.

The fourth section, west of Kilimanjaro, would comprise the southern part of Africa, from the Congo to the canal of Mozambique to the Cape of Good Hope, the southern Atlantic Ocean from Pernambuco and Lima, Bolivia, Paraguay, Uraguay, the Argentine Republic, Patagonia, the Fire Islands, the Malouine Islands, Sandwich and Shetland Islands, and the southern part of the Pacific Ocean east of the present 160th degree of latitude.

These would be the four sections, separated by the line of zero in calculating the sea-level changes. Now, the question was to indicate the effects produced on the surface of the four sections in consequence of the displacement of the oceans.

Upon each of these sections there was a central point on which the effect would be at a maximum, either by the oceans rising up or by the waters retiring entirely. The calculations of J.T. Maston had established without a doubt, that at each of these maxima points the greatest height obtained would be 8,415 metres. It was therefore certain that the consequences would be most severe against the security of those points through the operation carried out by Barbicane & Co. The two effects may be considered separate in their action.

In two of the sections situated opposite each other in the northern hemisphere and in the southern as well, the oceans would retreat and invade the two other sections, opposing each other in each of the two hemispheres.

In the first section: The Atlantic Ocean would be nearly entirely emptied and the maximum point of depression being nearly at the region of Bermuda, where the ground would appear, if the depth of the ocean was inferior at that point to 8,415 metres. Consequently between Europe and America vast territories would be discovered which the United States, England, France, Spain, and Portugal could claim according to the geographical situation, as these powers might wish to do. It must be observed that in consequence of the falling of the oceans the air will also fall equally as much. Therefore the barometric pressure of Europe and that of America will be modified to such an extent that cities, situated even 20 or 30 degrees from the maxima points would only have the quantity of air which is now actually found in a height of one league in the atmosphere. The principal cities, such as New York, Philadelphia, Charleston, Panama, Lisbon, Madrid, Paris, London, Edinburgh, Dublin, Cairo, Constantinople, Dantzig, Stockholm, on one side, and the cities corresponding in latitude on the other side, would keep their normal position with regard to the general level of the air. In regard to Bermuda, air would be missing there the same as it would be wanting to aeronauts who go higher than 8,000 metres. Therefore, it would be impossible to live there.

The same effect would obtain in the opposite section, which would contain the Indian Ocean, Australia, and a part of the Pacific Ocean, which would be thrown partly on the southern seacoasts of Australia.

The air into which they would be thrown would be very clear; there was no doubt on that point, but it would not be dense enough for human wants.

These in general were a part of the modifications which would take place in the two sections in which the oceans would be more or less emptied. There would undoubtedly appear new islands and mountains in such parts as the water did not entirely abandon.

But if the diminuation of the thickness of the air did not bring enough inconveniences to those parts of the new continents raised to the high zones of the atmosphere, what was to be the case of those parts which the errruption of waters put below the surface? We may still breathe under the diminished pressure of air below the atmospheric pressure. On the contrary, under a very few inches of water we cannot breathe at all, and this was the condition in which the other two sections found themselves. In the section northwest of Kilimanjaro the maximum point would be at Yakoutsk, in Siberia.

From this city submerged 8,415 metres under the water, less its present actual altitude, the liquid mass, decreasing, would extend to the neutral lines, drowning the greater part of Asiatic Russia and of India, of China, of Japan, and of American Alaska, to the Behring Sea. In regard to St. Petersburg and Moscow on one side, and Calcutta, Bangkok, Saigon, Pekin, Hong Kong and Yeddo on the other side, these cities would disappear under a cover of water sufficient to drown all Russians, Hindoos, Siamese, Cochin Chinese, Chinese and Japanese, if they did not have time to emigrate before the catastrophe.

In the section southeast of Kilimanjaro the disasters would be equally marked. This section is in a great part covered by the Atlantic and Pacific oceans, the level of which would raise 8,415 metres at the Archipelagos of the Azores. All this vast area would disappear under this artificial deluge, among others the angle of Southern Africa from Guinea and Kilimanjaro to the Cape of Good Hope, and the triangle of South America formed by Peru, Central Brazil, Chili, and the Argentine Republic, as far as Terra del Fuego and Cape Horn. The Patagonians, high as they are located, would not escape this immersion, and would not even have opportunity of taking refuge on that part of the Andes, as the highest points of that range would not be visible at all in this part of the globe.

This, then, must be the result, the lowering of the upper and raising of the lower sections, and an entirely new surface to the oceans, produced by the corruscations in the surface of the earth's sphere. Such were the happenings which would result, and against which the people of this world had no help if they could not promptly stop Barbicane & Co. in their criminal attempt.

CHAPTER XVI

CHAPTER XVI

IN WHICH A CROWD OF DISSATISFIED PEOPLE BREAK *INTO THE CELL OF J.T. MASTON.*

***

After this public notice there was nothing left but to wait for the coming danger or to run away to the neutral lines, where there would be no danger. The threatened people were, in general, divided into two classes – "the people who would be suffocated and those who would be drowned." This communication roused many different suggestions, which, however, all turned into the strongest and most violent protestations against the schemer and schemers. Among those who would suffocate were the Americans in the United States, the Europeans of France, England, Spain, etc. Even the prospect of annexing territories now at the bottom of the ocean was not sufficient to make them quietly accept these changes. Paris, carried towards the new pole a distance about equal to that which separates it now from the old one, would gain nothing by it. It would have a continued Spring, it is true, but it would lose considerable air. And this was not satisfactory to the Parisians, who like to have as much air as possible, and boulevard property and cafés went begging. Among those who would be drowned were the inhabitants of South America, of Australia, Canada, India, Zealand, etc. Great Britain would suffer the loss of her richest colonies, which Barbicane & Co. would take away from her through their operation. Evidently the Gulf of Mexico would constitute a vast kingdom of the Antilles, of which the Yankees and Mexicans could claim possession by the principles of the Monroe doctrine. The islands of the Philippines, Celebes and the water around them would leave vast territories of which the English and Spanish people could take possession. It is a vain compensation. It did not at all balance the loss due to the terrible flood.

If under the new oceans only Samoyedens, Lapons of Siberia, Feugans, Patogonians – even Tartars, Chinese, Japanese, or a few Argentines – would suffer and be lost, perhaps the civilized powers would have accepted this sacrifice complacently. But too many powers took part in the great catastrophe not to raise a torrent of protest.

And what especially concerned Europe was, that although the central part of it would be nearly intact, it would be raised in the west and lowered in the east, half suffocated on one side and half drowned on the other. This was not very acceptable. The Mediterranean Sea would be almost emptied, and this would not be very agreeable to the Frenchmen, Italians, Spaniards, Greeks, Turks, and Egyptians, who by their situation on the coast, had indisputable rights in ocean travel. And then, what good would be the Suez Canal, which would be saved by its position on the neutral line? But what use could be made of this immense work of Lesseps when there was no longer the Mediterranean on one side of the isthmus and the Red Sea on the other, at least, within any reasonable distance of it?

No, never, never would England consent to see Gibralter, Malta, and Cyprus transformed into mountain-tops, lost in the clouds, so that its men-of-war could no longer approach them. No, she would not be satisfied with the possession of some of the territory which would be gained from the Atlantic Ocean. Major Donellan had, however, prepared already to return to Europe to secure his rights on this new territory in case the operation of Barbicane & Co. should succeed. It is seen how protests came from all parts of the world, even from States where the changes would be imperceptible, because their people were interested in some other direction more or less.

These protestations became more and more violent after the arrival of the cablegram from Zanzibar which indicated the point of shooting, and which it was found necessary to publish the above report to explain. President Barbicane and Captain Nicholl as well as J.T. Maston, were put under the ban of humanity and declared outlaws. But what a business all this created for the newspapers. What sales they had, and how the circulations ran up; how on many occasions they were forced to print extra editions. It is perhaps the first time in journalistic history that they were all united with each other, as they generally quarrel incessantly. This was not a European or an American affair; it was an affair which concerned the whole world. It was like a bomb falling into a powder magazine.

In regard to Maston, it looked as if his last hour had come. A rabid crowd rushed into his prison on the evening of Sept. 17, with the intention of lynching him, and the jailer did not put any obstacles in their way. They rushed along the corridor but the cell of J.T. Maston was empty. Mrs. Evangelina Scorbitt had come to his help with a heavy purse of gold, and he had made his escape. The jailer had been bribed by an amount of money on which he could live the rest of his life without working. He remembered that Baltimore, Washington, New York, and many of the principal cities of America were on the line of those parts which would be raised, and which would still have enough air for the daily consumption of their inhabitants.

J.T. Maston had gained a quiet resting spot and a safe place from the enraged crowd of people, and so this great man owed his life to the devotion of a loving woman. There were only four days to wait, four days only before the gigantic operation of Barbicane & Co. would be accomplished. The public notice had been read far and wide and had created as much public excitement as such a momentous document only could. If there were at the beginning a few sceptics on the subject, there were none at present. The various governments had notified in haste those of their provinces which would be raised into the air and those, a much larger number, the territory of which would be overrun with water. In consequence of this advice sent by telegraph over the five continents of the world an emigration began such as had never been seen before. Every race was represented, white, black, brown, yellow, etc., in one chromatic procession. Unhappily, time was wanting for all to secure safety. The hours were now counted. A few months notice would be required for the Chinese to leave China, the Australians, Australia, the Siberians, Siberia. In some instances the danger was a local one as soon as the place of the shooting was known, so the fright became less general. Some provinces and even some States began to feel easy again. In a word, except in the regions directly threatened, there was only felt an apprehension of the terrible shock. And during all this time Alcide Pierdeux was saying to himself, "How in the wide world can President Barbicane make a cannon a million times larger than that of twenty-seven centimetre? This Maston, I would like very much to meet him – to have with him a talk upon this subject. This does not agree with anything sensible, it is too enormous and too improbable."

Be this as it may, the failure of the operation was the only hope which was left for certain parts of the world to escape more terrible destruction.

CHAPTER XVII

WHAT HAD BEEN DONE AT KILIMANJARO DURING EIGHT *MONTH OF THIS MEMORABLE YEAR.*

***

The country of Wamasai is situated in the eastern part of Central Africa, between the coast of Zanzibar and the regions of the large lakes, where the Victoria Nyanza and the Tanganiyka form a great interior ocean. The part best known is that which has been visited by the Englishman Johnston, Count Tekeli and the German doctor Meyer. This mountainous land is under the sovereignty of Sultan Bali-Bali, whose people consist of 30,000 or 40,000 Negroes.

Three degrees below the Equator is situated the chain of Kilimanjaro, which here reaches its greatest altitude. Among other peaks is the Mount of Kibo, with an altitude of 5,704 metres. The important ruler of this region has under his domination towards the south, north, and west the vast and fertile plains of Wamasai, which stretch from the lake of Victoria Nyanza across the province of Mozambique.

A few leagues below Kilimanjaro is the small village of Kisongo, the regular residence of the Sultan. This capital is in reality only a large hamlet. It is occupied by a very intelligent and industrious people, who work themselves as industriously as their slaves under the iron rule which Bali-Bali imposes on them.

This Sultan rightly ranked as one of the most remarkable rulers of those people of Central Africa who try to escape the influence, or more correctly the domination of England. At this capital of Kisongo, President Barbicane and Capt. Nicholl accompanied by six men who were devoted to them, arrived in the first week of January of the current year. On leaving the United States, whence their departure was only known to Mrs. Evangelina Scorbitt, and J.T. Maston, they had embarked in New York for the Cape of Good Hope, whence a vessel transported them to Zanzibar, on the island of the same name. There a bark secretly chartered by the Sultan brought them to the port of Mombas, on the African border on the other side of the channel. An escort sent by the Sultan waited for them at this port, and after a hard voyage nearly a hundred leagues across this terrible region, obstructed by forests, deep marshes, etc., they arrived at the royal residence. After knowing the calculations of J.T. Maston, President Barbicane had already put himself in communication with Bali-Bali through the help of a Swedish explorer, who had passed several years in this part of Africa. As the Sultan had become one of their most ardent admirers since their trip to the moon, a trip whose reputation had gone as far as these countries, he had a great friendship for these courageous Yankees. Without telling him for what purpose it was, Impey Barbicane had easily obtained permission from the Sultan to undertake important works at the southern foot of Kilimanjaro. In return for a large sum, estimated at $300,000, Bali-Bali had bound himself to furnish them all the workmen necessary. In other words, the captain and his friends were authorized to do at Kilimanjaro whatever they liked to do. They could dispose of the large chain of mountains according to their desires; they could tear them down if they liked, or they could take them away if they would be able to do so. In consequence of these arrangements, which the Sultan had made at his own figure, the North Polar Practical Association was as much proprietor of this country as they already were to the polar region. The reception which President Barbicane and Capt. Nicholl received at Kisongo was very cordial. Bali-Bali displayed an admiration amounting to adoration for these celebrated travellers who had made this dangerous voyage to reach the country around the North Pole.

He had in short an extraordinary sympathy for the creators of these mysterious operations which were going to be accomplished in his kingdom. He also promised them absolute secrecy on his part as well as on the part of his people, whose co-operation was assured to them. Not a single Negro who worked at their shop would be allowed to leave them for a single day under pain of the most severe punishment. This is how this operation was veiled in mystery so that the most active and sharpest agents of America and Europe failed to penetrate it. If it was finally discovered it must have been that the Sultan modified his severe rules after the accomplishment of the works and that there were traitors and babblers even amongst the Negroes. In this way Richard W. Trust, consul at Zanzibar, had received wind of what was going on at Kilimanjaro. But then at that date, the

13th of September, it was too late to stop President Barbicane in the accomplishment of his design.

And now, why had Barbicane & Co. chosen the Wamasai for the theatre of their operations? First, because the country suited them in regard to its geographical situation, as it was in a very little known part of Africa, and as it was very far from the territory usually visited by travellers. Then, the mass of Kilimanjaro offered them all the qualities of solidity and material necessary for their work. And, moreover, on the surface of this country were found the raw materials which they needed in a condition very easy to handle. A few months before leaving the United States President Barbicane had learnt from the Swedish explorer that at the foot of Kilimanjaro iron and coal were plentiful on the ground. No mines to dig into, no fields to explore a thousand feet deep in the earth's shell. Iron and coal were so plentiful even for this great undertaking that they only had to stoop down to pick it up. In other words, there existed in the neighborhood of this mountain enormous fields of nitrate of soda and of iron pyrites, which were necessary for the manufacture of melimelonite. President Barbicane and Capt.

Nicholl had taken with them only ten people, of whom they were absolutely sure, and no one else. These ten men had to supervise the 10,000 Negroes put at their disposal by Bali-Bali, and to them was given the task of manufacturing the monster cannon and its not less monster projectile. Two weeks after the arrival of President Barbicane and his associate at Wamasai three large workshops were established at the southern foot of Kilimanjaro, one for the cannon foundry, the second for the manufacture of the projectile, and the third for the manufacture of the melimelonite.

Now, first of all, how had Barbicane & Co. met the problem of manufacturing a cannon of such colossal dimensions? We will see and understand at the same time that the difficulty of creating such a device was not easily comprehensible by the inhabitants of the world. In reality the making of a cannon a million times larger than that of twenty-seven centimetres was a superhuman work. Already great difficulties had been met in the manufacture of pieces of forty-two centimetres long, which would throw projectiles of 780 kilos with 274 kilograms of powder. Barbicane & Co. did not think of these difficulties. It was not a cannon, not even a mortar, which they intended to make, but simply a gallery bored in the massive rock of Kilimanjaro, – a shaft of a mine, if you wish to call it so.

Evidently this shaft of a mine, this enormous elongated mine, could replace a metal cannon the fabrication of which would have been as dear as difficult and to which it would be necessary to give an unwieldy thickness to avoid all risk of an explosion. Barbicane & Co. had always entertained the idea of operating in this manner, and if the notebook of J.T. Maston mentioned a cannon it was that of 27 centimetres which had been used in the calculations as a basis. Consequently a spot was chosen at a height of a hundred feet on the southern slope of the chain. Nothing would be in the way of the projectile when it would fly out of the mouth of this tunnel bored in the massive rock of Kilimanjaro. It was with extreme precision and not without very hard work that the men could dig this gallery. But Barbicane & Co. could readily make perforations with simple machines put in action by means of compressed air which was secured by using the powerful falls of water from the mountains. In the holes bored through the headings of the shaft were placed charges of melimelonite. And nothing more was necessary than this violent explosive to shiver the rock, extremely hard as it was.

The thousands of workmen, led by their ten co-operators under the general direction of Barbicane & Co., labored with a great deal of zeal and intelligence to bring the work to a speedy end. At the end of six months the shaft measured 27 metres in diameter and the lining of it 6 metres in thickness. As it was absolutely necessary that the projectile should glide through a bore perfectly smooth the interior of it was covered with a casting exactly prepared. In reality this part of the work was very similar to that of the celebrated Columbiad, of Moon City, which had sent the projectile to the moon. But such work as this is impossible to the ordinary engineers of this world at present.

As soon as the boring was finished the workmen pushed on with the work at the second workshop.

At the same time that this metallic lining was being made they were also employed at making the enormous projectile. For this operation it was necessary to obtain a cylindrical mass which would weigh 80,000,000 kilograms, or 180,000 tons. It must be understood that there was never any idea of melting this projectile in one single piece. It had to be manufactured in thousand-ton pieces, which would be hoisted one after the other into the shaft and put in place over the chamber where the melimelonite was stored. After having been jointed each to the other, these pieces would form a compact whole, which would fit the sides of the tubular lining. In regard to the construction of the massive furnaces to effect the melting of the metal, there was met perhaps the greatest difficulty. Ten furnaces of ten metres each in height were at the end of a month in working order and able to produce each 180 tons per day. This would be 1,800 tons for twenty-four hours – 180,000 tons after 100 work-days.

In regard to the third workshop, made for the manufacture of the melimelonite, the work was easily done, but under such secret precautions, that the composition of this explosive it has not been possible to state perfectly. Everything went along splendidly. It could not have been possible to have met with more success in any factory. One would hardly expect to escape an accident of some sort on a three-hundred-thousand franc job. It is easily understood that the Sultan was delighted.

He followed the operation with indefatigable interest. And the presence of His Majesty helped greatly to make these Negroes work as hard as possible. One day Bali-Bali asked what all these operations were going on for. He received his reply from President Barbicane:

> “It is a work,” (said he,) “which will change the face of the earth – a work which will bring the greatest glory on the greatest Sultan of all the Eastern kings.”

By the 29th of August the works were entirely finished.

The shaft was bored to the wished-for point. It was provided with a smooth bore of six metres diameter. At the bottom of the shaft were placed the 2,000 tons of melimelonite; then came the projectile 105 metres long. After deducting the space occupied by the powder and projectile there remained still 492 metres before the muzzle was reached, which secured all the effect possible by the recoil produced by the expansion of the gas.

Now, the first question which might come up was, would the projectile deviate from the trajectory assigned to it by the calculations of J.T. Maston? In no way, for the calculations were absolutely correct. They indicated to what extent the projectile would deviate to the east of the meridian of Kilimanjaro because of the rotation of the earth on its axis, and what would be the form of the curve which it would describe because of its enormous initial velocity. Secondly, would it be visible during its course? No, because in going out of the shaft it would be thrown in the shadow of the earth and it could not be seen, for in consequence of its low trajectory it would have a very sharp angle of velocity compared with the earth's course. In fact, Barbicane and Capt. Nicholl could well be proud of their work, which had so far succeeded in its every detail. Why was J.T. Maston not there to watch this great operation, founded on the figuring which he had done? And who was it that kept him so far away, so very far, when this terrible detonation would wake the echoes as far as the furthermost horizon of Africa?

Thinking of him, his two associates did not know that the Secretary had been compelled to keep away from Ballistic Cottage after having got out of prison and hidden himself in a safe place away from the savage people. They did not know to what extent indignation had been roused against the engineer of the N.P.P.A. They did not know that they, too, would have been burnt or hanged and tortured to death if it had been possible to have reached them. Really, they ought to have been glad that at the moment when the shooting would take place they would only be saluted by the cries of this Negro people of Eastern Africa, "Well, at last!" said Capt. Nicholl to President Barbicane, when on the 22d of September they were standing before their finished work. "Yes, at last," said Impey Barbicane. "What a chance it was that placed at our disposition this admirable melimelonite!" said Capt. Nicholl. "Which will make you the most illustrious person on the earth, Nicholl." "Without doubt, Barbicane," modestly answered Capt. Nicholl. "But do you know how much it would have been necessary to dig out Kilimanjaro if we only had gun-cotton equal to that which threw our projectile to the moon?"

"How much, Nicholl?"

"One hundred and eighty galleries, Barbicane."

"Well, we would have digged them, Captain."

“And 180 projectiles of 180,000 tons.”

“We would have melted them, Nicholl.”

“It was useless to expect reasonable conversation between two persons of this type. But after they made the trip to the moon, what would they not be capable of? On the very same evening only a few hours before the minute when the gun was to be fired, and while President Barbicane and Capt. Nicholl were congratulating themselves, Alcide Pierdeux, closeted in his studio at Baltimore, uttered a cry of hurrah! as if he were crazy.

Then, suddenly getting up from the table, which was covered with figures and calculations, he cried out:

“Ah! What a fool Maston is! – what a stupid fellow! His whole problem will go in the soup! Christopher Columbus! Why did I not see this before? If I only knew where he was at this moment I would invite him to have supper with me and to sip a glass of champagne at the very moment when they are going to fire off the gun.”

And after these and many exclamations which he generally used in playing whist he said: "Oh, the old fool! Without a doubt he must have been dull when he made his calculations for this affair of Kilimanjaro. He will find it very necessary to make another. Oh, what a fool with his cannon!"

CHAPTER XVIII

CHAPTER XVIII

IN WHICH THE POPULATION OF WAMASAI ASSEMBLE *TO HEAR PRESIDENT BARBICANE SAY "FIRE" TO CAPT. NICHOLL.*

***

It was in the evening of the 22d of September, that memorable date which public opinion credited with an influence as unlucky as that of the 1st of January of the year 1000. Twelve hours after the sun had passed the meridian of Kilimanjaro, that is at midnight, Capt. Nicholl was to touch off the terrible cannon.

Kilimanjaro being 35 degrees east of the meridian of Paris, and Baltimore 79 degrees east of said meridian, there was a difference of 114 degrees between these two places, or 456 minutes in time, or 7 hours and 36 minutes. So the exact moment at which the shooting would take place would be 5 hours and 24 minutes post meridian in that great city of Maryland. The weather was magnificent.

The sun had just gone down on the plains of Wamasai, behind a horizon of perfect purity. It was impossible to wish for a prettier night, one more calm or starry, in which to throw the projectile across space. Not a cloud would be mixed with the artificial vapors developed by the deflagration of the melimelonite.

Who knows, perhaps President Barbicane and Capt. Nicholl regretted that they were not able to get into the projectile. In the first second they would have travelled 2,800 kilometres. Sultan Bali-Bali, with the great personages of his court, that is, his Finance Ministers and his Ministers of Public Works, together with the Black Brigade, who had helped in the great work, were all assembled to watch the different steps of the shooting.

But, with great precaution, they had all taken a position three kilometres from the shaft bored in the Kilimanjaro, so that they would have nothing to fear from the concussion of the air.

Several thousand natives, deputed from Kisongo and neighboring States in the south of the province, by the orders of the Sultan, were present to witness this splendid spectacle. A wire was stretched, connecting an electric battery to the touch-hole of the shaft, ready to send the current and start the deflagration of the melimelonite. As a preliminary an excellent meal had been served at the table of the Sultan for his American guests and the persons of his court, all at the expense of Bali-Bali, who did everything very grandly as long as he was reimbursed by the members of the firm of Barbicane & Co.

It was 11 o'clock when this feast, commenced at 7:30, was finished, and at the end of it the Sultan proposed a toast to the engineers of the N.P.P.A. and to the success of their great enterprise. An hour yet, and the change in the geographical and climatic conditions of the earth would be accomplished.

President Barbicane, his associate, and his ten helpers took their places around the cannon, to the interior of which ran the wire of the electric battery. Barbicane with his chronometer in his hand counted the minutes, and never in his life did they seem so long to him. The minutes seemed not merely years but centuries. At ten minutes before midnight Capt. Nicholl and Barbicane approached the key which put the electric thread in communication with the shaft of Kilimanjaro. The Sultan, his court and the crowd of natives formed an immense circle around the cannon. It was important that the shooting should take place at the exact moment indicated in the calculations of J.T. Maston – that is, at the moment when the sun would cut that equinoctial line which it would never leave again in its apparent orbit around the earth. Five minutes to twelve, four minutes, three minutes, two minutes, one minute to twelve–

President Barbicane watched the hands of his chronometer, lighted by a lantern which was held by one of his helpers, while Capt. Nicholl, his finger on the button of the apparatus, was ready to connect the circuit of electricity.

Twenty seconds, ten seconds, five seconds, one second. Not the slightest tremor could be noted in the hand of the impassive Nicholl. His partner and himself were no more excited than, at the moment when they waited, sitting in the projectile, for the Columbiad to fire them to the regions of the moon.

“Fire,” ordered President Barbicane.

At this moment Capt. Nicholl pressed the button. A terrible detonation followed, the echoes of which spread to the furthest corners of the province of Wamasai. A sharp whistle passed the crowd, a terrible rush of air, caused by the milliards of milliards of measures of gas, made by the instantaneous deflagrations of the 2,000 tons of melimelonite. It might be described as one of those meteors in which all the violence of nature is accumulated sweeping across the earth. The effect could not have been more terrible if all the cannons of the whole globe had been joined together with all the thunderbolts of heaven and all had united in one grand report.

CHAPTER XIX

IN WHICH J.T. MASTON REGRETS THAT THE CROWD DID NOT LYNCH HIM WHEN HE WAS IN PRISON.

***

The capitals of two worlds, the largest cities as well as the smaller ones, stood waiting terror-stricken. Thanks to the journals which had published the news broadcast over the world, every one knew the precise hour at which the shooting would take place and the local hour which corresponded with that of Kilimanjaro, situated 35 degrees east, allowing for the difference of longitude.

A few of the principal cities, the sun travelling a degree in four minutes were as follows:

At Paris, 9:40 P.M.

At St. Petersburg, 11:31 P.M.

At London, 9:30 P.M.

At Rome, 10:20 P.M.

At Madrid, 9:15 P.M.

AtBerlin, 11:20 P.M.

At Constantinople, 11:26 P.M.

At Calcutta, 3:04 A.M.

At Nanking, 5:05 A.M.

At Baltimore, it was said, twelve hours after the passage of the sun of the meridian of Kilimanjaro, it was 5:24 P.M. It is impossible to describe the pangs which were produced at this moment. The most powerful of modern pens would be helpless at the task. The people of Baltimore stood fearing that they would be swept off the surface of the earth by the terrible mass of water which would fall on their city. They expected to see the Bay of Chesapeake empty itself upon them. Then, besides, the city, even should the waters not come upon it, would be terribly shaken up by the shock which would be produced. The monuments would be destroyed; its best quarters swallowed up at the bottom of the abysses which would open through the surface of the ground. These fears ran through the different parts of the globe which were not scheduled for submersion by the upheaval of the oceans.

Every human being felt the marrow in his bones creep and shake at this fearful moment.

Yes, all trembled, all save one person, and that one was the engineer Alcide Pierdeux. As he had not had time to make known to the public the discoveries which he had made by means of his last calculations, he drank a bumper of champagne to the health of both worlds in the café of one of the best known hotels. Just as the twenty-fourth minute after 5 o'clock, corresponding with midnight at Kilimanjaro, was reached–

At Baltimore, nothing.

At London, Paris, Constantinople, Berlin, nothing, not the least shock.

Mr. John Milne, standing in his coal mine at Shamokui with a seismometer which he had arranged there, did not note the least abnormal movement in the earth's shell in this part of the globe. In Baltimore the heavens were cloudy and it was impossible to note in the apparent movement of the stars any derangement which would have indicated the change in the earth's axis.

What a night J.T. Maston passed in his place of safety which was unknown to all save Mrs. Evangelina Scorbitt! He was beside himself, this visionary engineer. He could not rest in his place of hiding. He seemed to have grown old in one day and looked sharply out to see if the daily course of the sun was modified. This would have been a certain proof of the success of his work. This change could not be seen even on the morning of the 23d of September, because at this date the star invariably rises in the east for all points of the globe. The next day the sun travelled over the horizon the same as it had always done.

The European delegates had assembled on the platform of their hotel. They had by their side instruments of extreme precision which would enable them to note if the sun took a course in the direction of the equator.

Well, nothing changed. A few minutes after the rising of the sun the great disc inclined away towards the Australian hemisphere. Nothing was changed in its apparent course.

Major Donellan and his associates saluted the heavenly torch with enthusiastic hurrahs, and gave it a reception like a favorite star in the theatre. The heavens were in superb condition, the horizon free from the vapors of the night, never did the great sun-god present a more beautiful aspect in such splendid condition before the astonished public. “And precisely at the place marked by the laws of astronomy,” said Eric Baldenak.

“Yes by our old astronomy,” said Boris Karkof, “and these fools pretended to destroy it.”

“Well, they will have their expenses to pay and ridicule to endure besides,” added Jacques Jansen, by whose voice Holland seemed to speak all alone.

“And the Arctic regions will eternally stay under the ice as they have discovered,” said Prof. Jan Harald.

“Hurrah for the sun,” said Major Donellan. “Such as it is, it has been and always will be sufficient for our earth.”

“Hurrah, hurrah,” repeated in single voice the representatives of old Europe. At this moment Dean Toodrink, who had not said anything so far, made this very cautious remark:

But perhaps they did not shoot yet.

“Not shoot yet,” said the Major. “Heaven grant that they have fired off the cannon twice rather than once.”

And that was exactly what J.T. Maston and Mrs. Evangelina Scorbitt were saying.

The wise and the ignorant were united this time by the logic of the situation. Even Alcide Pierdeux repeated it, and added: “Even if they did shoot, what is the difference? The earth will not stop waltzing on its old axis and turning as it used to do.”

In fact no one knew what had happened at Kilimanjaro. But at the close of the day an answer came to the question which was engrossing the attention of mankind.

A cablegram arrived in the United States, and here is what this dispatch, sent by Richard W. Trust, Consul at Zanzibar, contained:

"Zanzibar, Sept. 23, 7:25 A.M."

"To *John S. Wright, Minister of State:*

"The cannon was fired off yesterday evening at midnight exactly by the device bored in the southern part of Kilimanjaro. Passage of the projectile was accompanied with a powerful whirr and terrible detonation. Whole provinces destroyed by the concussion of the air. Ocean agitated as far as the Mozambique channel. A large number of vessels disabled and thrown on the coast. Towns and villages destroyed. Everything else is well.

"RICHARD W. TRUST."

Yes, everything else went on well. Nothing had been changed in the state of worldly affairs save the terrible disasters produced in Wamasai, which was partly deluged by the artificial waterspout, and the shipwrecks which were produced by the current of air. The same thing precisely happened when the Columbiad threw its projectile to the moon. The shock to the ground of Florida, was it not felt through a radius of 100 miles? Yes, certainly, but this time the effect should have been a hundred times stronger.

Whatever had happened the dispatch gave two pieces of information to the interested people of the old and new worlds.

> First – That the enormous cannon had been erected in the flank of Kilimanjaro.
>
> Second – That the gun had been fired at the fixed hour. And now, the whole world uttered an exclamation of intense satisfaction, followed by a great burst of laughter.

The trial which Barbicane & Co. had made had entirely failed. The calculations of J.T. Maston were good only for the waste basket. The N.P.P.A. could only announce its failure. But, perhaps, it might be that the secretary of the Gun Club had made a mistake in his calculations.

“Rather would I believe that I have been mistaken in the affection which I feel for him,” said Mrs. Evangelina Scorbitt.

But beyond all, the most discontented human being was J.T. Maston. When he saw that nothing had been changed in the movement of the earth, that the conditions remained precisely the same as they were since the creation, he hoped that some accident had prevented the success of Barbicane & Co., and that his associates had met with some disaster.

But there was the cablegram from Zanzibar which stated without a doubt that the operation had taken place.

Failed! ! And what of the formulas and calculations on which he had spent so much time? Is it possible that a cylinder 600 metres long, 27 metres wide, throwing a projectile of 180.000 kilograms, with the deflagration of 200 tons of melimelonite, with an initial velocity of 2,800 kilometres, would not be sufficient to move the earthly axis? It did not seem probable.

But why?

So J.T. Maston, in a state of violent excitement, declared that he would quit his retreat. Mrs. Evangelina Scorbitt tried in vain to prevent it. Not that she feared for his life, as all danger of that sort had passed. But the insults which he would have to bear, the jokes which would be cracked about him, the remarks which would be made in regard to his work – she wanted to spare him from these. And then, moreover, what would his associates of the Gun Club say? Did they not have to thank this man for the want of success of their operation and for making them ridiculous? Was he not the man who had figured out the whole affair and on whose shoulders rested all the responsibility?

J.T. Maston would not listen to any one. He resisted the begging and tears of Mrs. Evangelina Scorbitt. He went out of the house where he had kept himself hiding. He was recognized, and those who had trembled for fear of the consequences of his work now took revenge by joking and laughing at him, and this in many thousand different ways. He was forced to listen to jeering remarks, even from the street gamins. "Ah," they shouted, "here he is who wanted to change the axis of the earth, who wanted to discover coal mines around the

North Pole, who even wanted to remove it." In short, the Secretary of the Gun Club was compelled to return to the mansion of Mrs. Evangelina Scorbitt, who used all her wealth of tenderness to console him. It was in vain, however. J.T. Maston could not be consoled, as his cannon had produced upon the earth's sphere no more effect than a simple popgun would have done. A fortnight went by in this way, and the world resumed its daily routine and did not even think any longer of the projects of the N.P.P.A.

A fortnight and no news yet from President Barbicane and Capt. Nicholl. Had they perished by the discharge in the land of Wamasai? Had they sacrificed their lives in the most mysterious operation of modern times? No.

After the detonation both were overthrown along with the Sultan arid his court, and a thousand natives in one grand tumble, but they all got up after a little time strong and hearty.

"Did you succeed?" asked Bali-Bali, rubbing his shoulder.

"Do you doubt it?"

"Me doubt it?"

"But when will you know?"

"In a few days," said Barbicane.

Did he appreciate that the operation had failed? Perhaps. But he never would have acknowledged it before the Sultan at Wamasai.

Forty-eight hours later the two partners had taken leave of Wamasai, not, however, before having paid an enormous sum for the damage done to the country. As this amount of money went into the private purse of the Sultan, and as his subjects did not receive one cent of it, he had no reason to complain of the operation.

Then the two associates, followed by their ten helpers, reached Zanzibar, where they found a vessel to take them to Suez. From there under false names the steamer Morris brought them to Marseilles; then they took the train to Paris, where they arrived without having had any collision or accident, and taking the railroad to Havre they arrived in time to go to America by the Bourgogne of the Transatlantic line. In twenty-two days they made the trip from Wamasai to New York, and on the 15th day of October the two knocked at the door of the mansion of New Park, at three minutes past noon. An instant afterwards they found themselves in the presence of Mrs. Evangelina Scorbitt and J.T. Maston.

CHAPTER XX

IN WHICH THIS STORY, AS TRUTHFUL AS IT IS IMPROBABLE, *IS FINISHED.*

***

"Barbicane!!! Nicholl!!"

"Maston."

"You."

"We."

And in this plural pronoun, uttered simultaneously by the two associates in a single voice, might be heard a flood of irony and reproaches.

J.T. Maston pressed his iron hook on his forehead. Then, with a voice which seemed to stick in his throat, he said:

"Did your shaft at Kilimanjaro really have a diameter of twenty-seven metres?"

"Yes, sir."

"Did your projectile really weigh 180,000,000 of kilograms?"

"Yes."

"And was the shooting really done with 2.000 pounds of melimelonite?"

"Yes."

This thrice-repeated "yes" fell on J.T. Maston like masses of stone on his head.

"Then I can only conclude" – said he.

"What?" asked President Barbicane.

"As follows," said J.T. Maston. "As the operation did not succeed, the powder did not give to the projectile an initial velocity of 2,800 kilometres."

“Really?” said Capt. Nicholl, with a tone of sarcasm.

“Yes, your melimelonite is good only to charge pistols of straw.”

Capt. Nicholl sprang up at this remark, which was an outrageous insult to him.

“Maston!” said he.

“Nicholl!”

“You ought to be blown up with the melimelonite.”

“No, gun cotton; that is more sure.”

Mrs. Evangelina Scorbitt had to interfere and cool these two enraged gunners down.

“Gentlemen,” said she, between associates.

“And anyhow,” President Barbicane resumed, with a very calm expression, “what is the good of criminations? It is certain that the calculations of our friend, J.T. Maston, were correct, as it is certain that the explosive of our friend Nicholl had sufficient power. Yes, we have only employed known quantities of science. We lacked experience. Why did we fail? Perhaps we may never know."

"Well," said the Secretary of the Gun Club, "we will commence all over again." "And the money then which has been spent for this operation is a dead loss," observed Capt. Nicholl.

"But public opinion," said Evangelina Scorbitt, "would not allow you a second trial."

"What will become of our Arctic region?" said Capt. Nicholl." "Where will the stock of the N.P.P.A. fall to?" said President Barbicane. Well, it had already fallen so far that the stock was offered at the price of old paper.

This, then, was the result of the gigantic operation. This was the memorable fiasco to which the superhuman projects of Barbicane & Co. had led.

If ever engineers, unlucky engineers were laughed at in public, if ever the newspaper made drawings, songs, and paragraphs not at all flattering to the people mentioned in them, this occasion exceeded them all. President Barbicane, the Directors of the new Society and their associates of the Gun Club were universally sneered at. In every language they were made ridiculous, and to make it easier to the whole population of the world to read the scornful articles were printed in "Volapuk." In Europe, especially, all the remarks and songs to make the persons of the N.P.P.A. ridiculous were spread broadcast. The greatest hit was made by a Frenchman, who composed a ballad which was sung in every concert hall of France and America. But will we never know to what the failure of this enterprise was due? Did this failure prove that the operation was impossible of realization; that the powers at the disposal of mankind would never be sufficient to bring about a change in the earth's movement? Did it prove that the country around the North Pole would never be removed to those regions where the sun and heat would melt the ice without human help?

Information on this subject came a few days after the return of President Barbicane and Capt. Nicholl to the United States. A very simple paragraph appeared in the *Times* of the 17th of October. Here is the article:

> "We all know that the result of the operation to create a new axis has been nothing. However, the calculations of J.T. Maston, founded on established facts, would have produced the desired result if through an unexplainable slip an error had not been embraced in them from the beginning. When the celebrated secretary of the Gun Club took for a basis of his calculations the circumference of the earth's sphere, he figured it at 40,000 metres in place of 40,000,000 metres, and to which the failing of the operation is due.
>
> "Where could such an error come from? Who could have provoked it? . . . How could such a remarkable calculator commit such an error?
>
> "It is certain that had the problem of the modification of the earth's axis been correctly figured, it would have had been exactly solved. But this forgetting of three zeros has made a change at the end of the calculation of twelve naughts.

"It is not a cannon one million times larger than that of twenty-seven centimetres, which was necessary. A trillion of these cannons throwing a trillion projectiles of 80,000 tons each would be necessary to displace the North Pole, admitting that the melimelonite had the expansive power which had been attributed to it by Capt. Nicholl.

"Therefore the whole shock under the conditions under which it was produced has displaced the North Pole only three-thousandths of a milimeter, and has only changed the level of the ocean at the most nine-thousandths of three-thousandths of a milimetre. In regard to the projectile fired, it will be a small planet, and will belong in future to the solar system, sustained by solar attraction.

"ALCIDE PIERDEUX".

So this want of attention on the part of J.T. Maston at the beginning of his calculations had produced such a humiliating result for his Company.

But even if his associates were very angry against him, if everybody laughed and joked at him, it is only fair to state in his favor that this mistake which had wrecked the operation had spared the world a terrible catastrophe.

A flood of telegrams and letters arrived from all parts of the world congratulating J.T. Maston on his mistake of three naughts. J.T. Maston, more downhearted and crushed than ever, would not listen to the hurrahs which the world now uttered for him. President Barbicane, Capt. Nicholl, Tom Hunter, with wooden legs; Col. Bloomsberry, the gay Bilsby, and his associates would never pardon him. But Mrs. Evangelina Scorbitt she could not be angry with him, most excellent lady.

J.T. Maston had begun to do his calculations over again, refusing to admit that he was wrong at that point.

He was, however; the Engineer Alcide Pierdeux had not made a mistake. Having learnt his error at the last moment, when he had no time to make it known, he had remained perfectly composed among all the fright and terror of those about him. That was why he proposed a toast in champagne at the moment when the shooting was taking place in the Old World. Yes, indeed, three naughts had been forgotten in the circumference of the earth. Suddenly J.T. Maston remembered the whole matter.

It was at the beginning of his work when he had shut himself up in the “Ballistic Cottage,” and written the number 40,000, 000 on his blackboard. At that moment the electric bell began to ring with great force. J.T. Maston went to the phone. He exchanged a few words with Mrs. Evangelina Scorbitt. Suddenly a terrible stroke of lightning from the storm through the telephone knocked over his blackboard and himself. He got up, commenced to write over again the numbers which had been half rubbed out on his blackboard. He had just written the numbers 40,000 when the telephone rang for the second time. He went again to listen to Mme. Scorbitt, and when he did begin his work he forgot to put on the last three naughts of the earth's circumference.

It was the fault of Mrs. Evangelina Scorbitt. If she had not interrupted him he would not have been thrown on the floor by the shock from the telephone. He would not have noticed anything of lightning and thunder, and all his mass of figures and calculations would not have ended in a mistake.

What a terrible blow it was to this unhappy lady when J.T. Maston was compelled to tell her the circumstances which had produced the mistake! Yes, she was the cause of the disaster. It was on her account that J.T. Maston found himself dishonored through the long years which he bad yet to live, as nearly every member of the Gun Club usually lived to the age of a hundred years.

After this conversation at New Park, J.T. Maston had gone away from the mansion. He went back to his Ballistic Cottage and walked into his study muttering to himself: “Well, now I am not good for anything any more in this world.”

“Not even good enough to get married,” said a broken voice at his elbow.

It was that of Mrs. Evangelina Scorbitt. Absolutely crushed and heart-broken, she had followed him.

“Dear Maston” – she began.

“Well, yes,” said he, “but only under one condition – that I shall never make any mathematical calculations.”

“My dear friend, I have a horror of them,” answered the excellent widow.

Thus it happened that the Secretary of the Gun Club made Mrs. Evangelina Scorbitt Mrs. J.T. Maston.

In regard to the article of Alcide Pierdeux, we might say that it brought him into great celebrity and reputation.

It was translated into all languages, printed in every paper, and thus his name became known all over the world. The father of his old sweetheart had refused him his daughter's hand, after telling him that he could not give him his daughter, as he was too smart. But now, after having read this article and being unable to understand it without any help, he began to feel sorry and know better. He sent him an invitation to dine with him and his daughter.

CHAPTER XXI

VERY SHORT, SINCE ENOUGH HAS BEEN SAID TO MAKE THE WORLD'S POPULATION FEEL PERFECTLY SURE AGAIN.

***

And now the inhabitants of the world could again be perfectly easy. President Barbicane and Capt. Nicholl will not again begin that enterprise so woefully miscarried, J.T. Maston will not again figure out any calculations, however free from mistakes. The article of Alcide Pierdeux has told the truth. What the law of mechanics proves to us is that to produce a displacement of the axis of 23 degrees and 28 minutes, even with the melimel-onite, a trillion cannons similar to the one which had been bored into the cliff of Kilimanjaro would be necessary. But our whole sphere, bored over its whole surface, is too small to accommodate them. Therefore the inhabitants of the earth may sleep in peace. To modify the conditions in which the earth is moving is beyond the efforts of human-

ity. It is not meet that mere humanity should change anything in the order established by our Creator in the system of the universe.

Books For ALL Kinds of Readers

At ReadHowYouWant we understand that one size does not fit all types of readers. Our innovative, patent pending technology allows us to design new formats to make reading easier and more enjoyable for you. This helps improve your speed of reading and your comprehension. Our EasyRead printed books have been optimized to improve word recognition, ease eye tracking by adjusting word and line spacing as well as minimizing hyphenation. Our EasyRead SuperLarge editions have been developed to make reading easier and more accessible for vision-impaired readers. We offer Braille and DAISY formats of our books and all popular E-Book formats.